THE REGULATION OF PROLIFERATION AND DIFFERENTIATION IN NORMAL AND NEOPLASTIC CELLS

BRISTOL-MYERS CANCER SYMPOSIA

Series Editor
STEPHEN K. CARTER*
Science and Technology Group
Bristol-Myers Company

1. Harris Busch, Stanley T. Crooke, and Yerach Daskal (Editors).
Effects of Drugs on the Cell Nucleus, 1979.
2. Alan C. Sartorelli, John S. Lazo, and Joseph R. Bertino (Editors).
Molecular Actions and Targets for Cancer Chemotherapeutic Agents, 1981.
3. Saul A. Rosenberg and Henry S. Kaplan (Editors).
Malignant Lymphomas: Etiology, Immunology, Pathology, Treatment, 1982.
4. Albert H. Owens, Jr., Donald S. Coffey, and Stephen B. Baylin (Editors).
Tumor Cell Heterogeneity: Origins and Implications, 1982.
5. Janet D. Rowley and John E. Ultmann (Editors).
Chromosomes and Cancer: From Molecules to Man, 1983.
6. Umberto Veronesi and Gianni Bonadonna (Editors).
Clinical Trials in Cancer Medicine: Past Achievements and Future Prospects, 1985.
7. Paul A. Marks (Editor).
Genetics, Cell Differentiation, and Cancer, 1985.
8. Kenneth R. Harrap and Thomas A. Connors (Editors).
New Avenues in Developmental Cancer Chemotherapy, 1987.
9. Paul V. Woolley III and Kenneth D. Tew (Editors).
Mechanisms of Drug Resistance in Neoplastic Cells, 1988.
10. Emil Frei III (Editor).
The Regulation of Proliferation and Differentiation in Normal and Neoplastic Cells, 1989.

* *Series Editor for Volumes 1–8 was Maxwell Gordon.*

THE REGULATION OF PROLIFERATION AND DIFFERENTIATION IN NORMAL AND NEOPLASTIC CELLS

Edited by

EMIL FREI III
Dana-Farber Cancer Institute
Boston, Massachusetts

ACADEMIC PRESS, INC.
Harcourt Brace Jovanovich, Publishers
San Diego New York Berkeley Boston
London Sydney Tokyo Toronto

Academic Press, Inc.
San Diego, California 92101

United Kingdom Edition published by
Academic Press Limited
24–28 Oval Road, London NW1 7DX

Library of Congress Cataloging in Publication Data

The Regulation of proliferation and differentiation in normal and neoplastic cells / edited by Emil Frei III.
p. cm. -- (Bristol-Myers cancer symposia ; 10)
Proceedings of the 10th Bristol-Myers Symposium on Cancer Research, held in Boston, Mass., 1987.
Includes bibliographies and index.
ISBN 0-12-266970-3 (alk. paper)
1. Cancer cells--Growth--Regulation--Congresses. 2. Cell proliferation--Congresses. 3. Cell differentiation--Congresses. I. Frei, Emil, Date. II. Bristol-Myers Symposium on Cancer Research (10th : 1987 : Boston, Mass.) III. Series.
[DNLM: 1. Cell Differentiation--congresses. 2. Cell Division--congresses. 3. Cell Transformation, Neoplastic--congresses. W3 BR429 v. 10 / QZ 202 R3443 1987]
RC268.5.R449 1989
616.99'4--dc19
DNLM/DLC
for Library of Congress 88-37147
CIP

Printed in the United States of America
89 90 91 92 9 8 7 6 5 4 3 2 1

Contents

3 The Platelet-Derived Growth Factor: History, Chemistry, and Molecular Biology

CHARLES D. STILES

PART II CELL PROLIFERATION AND DIFFERENTIATION

4 Hemopoietic Growth Factor Action

E. RICHARD STANLEY

5 Lymphocyte Activation: Role of Cell Adhesion Molecules

P. ANDERSON, C. MORIMOTO, and S. F. SCHLOSSMAN

Contributors

Numbers in parentheses indicate the pages on which the authors' contributions begin.

P. ANDERSON (79), Division of Tumor Immunology, Dana-Farber Cancer Institute, Boston, Massachusetts 02115

CHARLES C. BASCOM (89), Departments of Cell Biology and Medicine, Vanderbilt University School of Medicine, Nashville, Tennessee 37232

ROBERT J. COFFEY, JR. (89), Departments of Cell Biology and Medicine, Vanderbilt University School of Medicine, Nashville, Tennessee 37232

MANUEL DIAZ (125), Departments of Medicine and Pathology, University of Chicago, Chicago, Illinois 60637

JEFF L. ELLSWORTH (197), Department of Biological Sciences, Stanford University, Stanford, California 94305

MARY GONZATTI-HACES (143), BRI-Basic Research Program, Frederick, Maryland 21701

JEAN M. GUDAS (3), Department of Biological Chemistry and Molecular Pharmacology, Harvard Medical School, and Division of Cell Growth and Regulation, Dana-Farber Cancer Institute, Boston, Massachusetts 02115

WILLIAM A. HASELTINE (165), Department of Pathology, Harvard Medical School, Boston, Massachusetts 02115

IRA HERSKOWITZ (21), Department of Biochemistry and Biophysics, University of California, San Francisco, California 94143

CYNTHIA HOY (197), Department of Biological Sciences, Stanford University, Stanford, California 94305

ANAND IYER (143), BRI-Basic Research Program, Frederick, Maryland 21701

GLENN B. KNIGHT (3), Department of Biological Chemistry and Molecular Pharmacology, Harvard Medical School, and Division of Cell Growth and Regulation, Dana-Farber Cancer Institute, Boston, Massachusetts 02115

MARIA M. KONARSKA (115), Center for Cancer Research and Department of Biology, Massachusetts Institute of Technology, Cambridge, Massachusetts 02139

RUSSETTE M. LYONS (89), Departments of Cell Biology and Medicine, Vanderbilt University School of Medicine, Nashville, Tennessee 37232

TIMOTHY McKEITHAN (125), Departments of Medicine and Pathology, University of Chicago, Chicago, Illinois 60637

C. MORIMOTO (79), Division of Tumor Immunology, Dana-Farber Cancer Institute, Boston, Massachusetts 02115

HAROLD L. MOSES (89), Departments of Cell Biology and Medicine, Vanderbilt University School of Medicine, Nashville, Tennessee 37232

MARIANNE OSKARSSON (143), BRI-Basic Research Program, Frederick, Maryland 21701

ARTHUR B. PARDEE (3), Department of Biological Chemistry and Molecular Pharmacology, Harvard Medical School, and Division of Cell Growth and Regulation, Dana-Farber Cancer Institute, Boston, Massachusetts 02115

MORAG PARK (143), BRI-Basic Research Program, Frederick, Maryland 21701

RICHARD S. PAULES (143), BRI-Basic Research Program, Frederick, Maryland 21701

FRIEDRICH PROPST (143), BRI-Basic Research Program, Frederick, Maryland 21701

JANET D. ROWLEY (125), Departments of Medicine and Pathology, University of Chicago, Chicago, Illinois 60637
NORIYUKI SAGATA (143), BRI-Basic Research Program, Frederick, Maryland 21701
ROBERT T. SCHIMKE (197), Department of Biological Sciences, Stanford University, Stanford, California 94305
S. F. SCHLOSSMAN (79), Division of Tumor Immunology, Dana-Farber Cancer Institute, Boston, Massachusetts 02115
R. IVAN SCHUMACHER (197), Department of Biological Sciences, Stanford University, Stanford, California 94305
PHILLIP A. SHARP (115), Center for Cancer Research and Department of Biology, Massachusetts Institute of Technology, Cambridge, Massachusetts 02139
STEVEN W. SHERWOOD (197), Department of Biological Sciences, Stanford University, Stanford, California 94305
NANCY J. SIPES (89), Departments of Cell Biology and Medicine, Vanderbilt University School of Medicine, Nashville, Tennessee 37232
JOSEPH SODROSKI (165), Department of Pathology, Harvard Medical School, Boston, Massachusetts 02115
SAMUEL H. SPECK (105), Dana-Farber Cancer Institute, Harvard Medical School, Boston, Massachusetts 02115
E. RICHARD STANLEY (51), Department of Developmental Biology and Cancer, The Albert Einstein College of Medicine, Bronx, New York 10461
CHARLES D. STILES (39), Department of Microbiology and Molecular Genetics, Harvard Medical School, and Dana-Farber Cancer Institute, Boston, Massachusetts 02115
JACK L. STROMINGER (105), Dana-Farber Cancer Institute, Harvard Medical School, Boston, Massachusetts 02115
ERNEST TERWILLIGER (165), Department of Pathology, Harvard Medical School, Boston, Massachusetts 02115
JOSEPH R. TESTA (143), BRI-Basic Research Program, Frederick, Maryland 21701
GEORGE F. VANDE WOUDE (143), BRI-Basic Research Program, Frederick, Maryland 21701

Editor's Foreword

Cancer is a disease of devastating impact for which prevention, diagnosis, and therapy have been developed within the pragmatic context of the fact that the basic origins of cancer at the cellular and molecular level remain unknown. At the current time, basic research on the cancer problem represents an exciting arena of ferment based on new understandings. The challenge is to take the new scientific findings and create a developmental therapeutic strategy for a more effective clinical attack on cancer.

The Tenth Bristol-Myers Symposium on "The Regulation of Proliferation and Differentiation in Normal and Neoplastic Cells" examined several approaches to the cancer problem evolving out of an increased basic scientific understanding. These approaches included work on growth factors and their mechanisms, molecular genetics, the study of oncogenes, increased understanding of the immune system, and elucidation of the mechanisms of drug resistance.

The Symposium was divided into two parts. On the first day the topic was cell proliferation and control mechanisms. The focus was on the cycle of cell division and how this process becomes aberrant in cancer cells. The topic for the second day was cell proliferation and differentiation; we elucidated new findings on growth factors and their relationship to the neoplastic process.

It is hoped that symposia such as this one will help foster the critical interactions necessary to achieve the fusion of the emerging new knowledge into effective therapeutic strategies, which will lead to an even more rapid improvement of the long-term disease-free survival rates for cancer patients.

Stephen K. Carter

Foreword

The outlook for long-term survival and even cure of many cancer patients is more positive today than at any other time in history. Yet despite all the great advances in cancer research and treatment, and in the face of new technologies and therapies now available, fundamental scientific questions overshadow the breakthroughs: Just what is the fine line between a normal cell and a malignant one? What mechanisms regulate that differentiation?

Cancer research experts around the world are dedicated to answering these questions. Their efforts focus on studying how cells divide and mature and, in particular, on identifying the genetic and biochemical similarities and differences between these processes in normal and malignant cells.

Important new findings resulting from this research were the focus of the tenth annual Bristol-Myers Symposium on Cancer Research, "The Regulation of Proliferation and Differentiation in Normal and Neoplastic Cells." The 2-day event was organized by the Dana-Farber Cancer Institute, chaired by Emil Frei III, and held in Boston on December 10 and 11, 1987.

The findings presented in this volume include new information on the cycle of cell division and recent research on the various genes, chromosome abnormalities, and viruses that cause a normal cell to go awry. Additional presentations explore the role of growth factors in causing and potentially curing cancer. The material also provides an in-depth look at the transformed cell that has crossed the line from normal to malignant.

Bristol-Myers Company has, for over a decade, supported such research through no-strings-attached funding of leading cancer research institutions. That funding now exceeds $12 million and has provided 25 grants to 23 cancer centers in the United States and abroad.

In addition, each spring we present the annual Bristol-Myers Award for Distinguished Achievement in Cancer Research in recognition of outstanding individual research. The recipient is selected by an independent peer review committee. In 1989, the 12th annual award was presented to Dr. Peter K. Vogt of the University of Southern California for his breakthrough discoveries of the *src* and *jun* oncogenes as well as his pioneering retrovirus research.

To help ensure the continued flow of information gained from cancer research, Bristol-Myers Company sponsors major scientific symposia organized by cancer research centers participating in the unrestricted grants program. Each of these has resulted in a published volume.

It is our hope that the information presented in this, the tenth such volume in this series, enlightens and broadens the reader's perspective and contributes in some measure to the exchange of ideas that fuels new research.

Richard L. Gelb
Chairman of the Board
Bristol-Myers Company

Preface

It has long been evident that the central feature of cancer is inadequately regulated cell proliferation. A closely related phenomenon is evidence for a block in cell differentiation. These two phenomena—cell proliferation and differentiation—are fundamental to all biology. Thus basic cancer research requires an understanding of normal cell proliferation and differentiation, and understanding cancer will be definitive when we appreciate at a molecular level the perturbations that occur in transformed cells with respect to proliferation and differentiation.

The tenth annual Bristol-Myers conference was initially conceived as one that would cover both clinical and basic aspects of proliferation and differentiation. Many cancer chemotherapeutic agents target on mechanisms essential to cell proliferation. However, in planning the program, it quickly became evident that both basic and clinical aspects of these phenomena could not be covered in a 2-day meeting. In view of the extraordinary advances in basic research in this area and the increasingly coherent body of knowledge across disciplines relating to proliferation and differentiation, it was decided to assemble some of the outstanding contributors representing diverse disciplines.

The first portion of the book covers cell proliferation and differentiation with emphasis on the normal cell. The latter portion deals with these processes within the transformed cell. The book will be of major interest to tumor biologists and will provide the clinical investigator with an important resource of basic information.

Emil Frei III

PART I

Cell Proliferation and Control Mechanisms

1

The Cell Cycle and Restriction Point Control

JEAN M. GUDAS, GLENN B. KNIGHT, AND ARTHUR B. PARDEE
Department of Biological Chemistry and Molecular Pharmacology
Harvard Medical School
and Division of Cell Growth and Regulation
Dana-Farber Cancer Institute
Boston, Massachusetts

I. Introduction

In normal mammalian cells, proliferation is a stringently regulated process culminating in the appearance of pairs of daughter cells at

THE REGULATION OF PROLIFERATION AND DIFFERENTIATION IN NORMAL AND NEOPLASTIC CELLS

division. There is an emerging consensus that exogenous growth factors regulate the proliferation of normal cells and that this regulation is deranged in cancer cells. Specific growth factors, such as platelet-derived growth factor (PDGF), stimulate fibroblast cells to emerge from quiescence (G_0) and become competent to enter the cell cycle (Pledger *et al.,* 1977, 1978). Other growth factors, including epidermal growth factor (EGF) and insulin or somatomedin C (IGF-1), are then required for progression of cells through the G_1 phase (Leof *et al.,* 1982; Campisi and Pardee, 1984). These growth factors act during this period to influence transcriptional, posttranscriptional, and biochemical events that eventually prepare the cells for DNA synthesis and mitosis. Related to this, Greenberg and Ziff (1984) have shown that a battery of genes has increased transcription rates at programmed intervals during the cell cycle. The transcriptional response of c-*fos* occurs within 15 min of PDGF stimulation with c-*myc* transcription following within 1–2 hr (Greenberg and Ziff, 1984; Müller *et al.,* 1984; Kruijer *et al.,* 1984). From the V point, which corresponds to the entrance point of cycling cells into G_1 from M, only IGF-1 is required (Yang and Pardee, 1986). Subsequent to G_1, growth factors are no longer required, as cells deprived of all factors will continue through S, G_2, and M until they again reach G_1.

Specific events occurring in the G_0–G_1 part of the cell cycle are controlled, first at emergence from quiescence, and then later in G_1 at the restriction point prior to the beginning of S phase. Deranged control of biochemical processes that occur at these control points are thought to result in the relaxed growth factor requirements of transformed and tumorigenic cells. This relaxed requirement for growth factors may result from deregulation of genes in pathways normally activated only when growth factor receptors are occupied (Ferrari and Baserga, 1987). In this chapter we will be concerned with events that occur at the second controlling point in G_1, just prior to DNA synthesis.

II. Restriction Point Control

Several lines of evidence indicate that there is a regulated event in late G_1, the restriction (R) point, occurring approximately 2–3 hr prior to the onset of DNA synthesis (Pardee, 1987). If serum is removed from exponentially growing cells, they continue to enter S phase for about 3 hr, after which time no more cells progress (Campisi *et al.*, 1982). These observations suggest that cells which had not passed the R point could not enter S phase. Thus, cells become independent of growth factors in the 2–3 hr preceding S phase. Furthermore, concentrations of cycloheximide which only inhibit total protein synthesis by 50–70% dramatically affect the pre-R point part of the cell cycle, causing an increase in cycle time of about 4-fold (Rossow *et al.*, 1979). Some process requiring rapid protein synthesis is necessary for allowing transit of cells past the R point. When protein synthesis is blocked for increasing lengths of time with a high dose of cycloheximide, the time after its removal that is required for cells to enter S phase is in excess of the inhibitory period (Campisi and Pardee, 1984). These findings show that protein synthesis is required prior to the R point and suggest that some critical unstable protein must accumulate to a threshold level for cells to continue into S phase. In order for this protein to accumulate, rapid synthesis is required to outdistance degradation. From the excess delay data, a half-life of about 2.5 hr was calculated for this putative labile protein (Campisi *et al.*, 1982; Campisi and Pardee, 1984). Further evidence to support the idea that this phenomenon is related to growth control comes from the finding that transformed cells do not have an excess delay for entry into S when treated in the same way with cycloheximide (Campisi and Pardee, 1984). Presumably, transformed cells contain sufficient quantities of the required labile protein from either stabilization or increased synthesis.

In an effort to define proteins fulfilling the requirements of a regulatory factor, a series of 2-D gel analyses of cell proteins at different times during G_1 was performed (Croy and Pardee, 1983).

One protein spot, corresponding to a 68-kDa protein having a pI of 6.3, was observed to appear near the R point and to disappear within 3 hr in the absence of protein synthesis. Moreover, in transformed cells this protein was found to be quite stable and to be present in quiescent populations. We are presently attempting to purify and characterize this putative regulatory protein in order to ascertain its function in cell cycle control at the R point.

III. Events Involved in the Onset of DNA Synthesis

A. DNA Biosynthetic Enzymes as Markers for S Phase

The sequence of events leading to the onset of DNA synthesis is extremely complex, involving a variety of pathways of signal transduction from the cell's exterior to the nucleus. This, coupled with the complexity of the DNA replication machinery, makes resolution of the sequence of events a difficult problem. Concomitant with the onset of DNA synthesis, however, the activities of a number of enzymes involved in DNA synthesis, as well as histone synthesis, rise sharply. These enzymes include thymidine kinase (TK) (Stubblefield and Mueller, 1965; Johnson *et al.*, 1982; Coppock and Pardee, 1985; Yang and Pardee, 1986), thymidylate synthase (TS) (Navalgund *et al.*, 1980; Coppock and Pardee, 1985; Yang and Pardee, 1986), and dihydrofolate reductase (Johnson *et al.*, 1978; Hendrickson *et al.*, 1980; Leys and Kellems, 1981). TK does not trigger the onset of DNA synthesis, since mutants defective for TK make DNA at the appropriate time. Furthermore, this increase is not a consequence of DNA synthesis because the change in TK (and TS) activity is unaffected by blocking DNA synthesis with inhibitors (Navalgund *et al.*, 1980; Johnson *et al.*, 1982; Coppock and Pardee, 1985). Thus, controlling mechanisms set in place by the R point induce a concerted production of histones, DNA biosynthetic enzymes, and DNA. Since the induction of DNA biosynthetic enzymes parallels the onset of DNA replication, the endpoint of TK induction can be equated with DNA synthesis (Coppock and Pardee, 1985; Yang and Pardee, 1986). Using this approach, one can investigate the molecular mechanisms underlying TK induction and then work backward through the sequence of preceding events to define the more global controlling factors.

B. Regulation of Thymidine Kinase Enzyme Activity

In mammalian cells there are two nonallelic forms of thymidine kinase: one isozyme resides in the cytosol and the other in the mitochondria (Kit, 1976). The cytosolic enzyme is of particular interest as its activity has been shown to closely parallel the growth rate of cells. Rapidly proliferating fetal and neoplastic tissues, as well as exponentially growing cells in culture, express high levels of cytosolic TK. However, slowly growing, resting, or terminally differentiated cells have very low levels (Claycomb, 1979; Merrill *et al.*, 1984b). We have previously shown that TK activity is almost undetectable in quiescent BALB/c 3T3 cells and that it increases dramatically about 12–14 hr after restimulation with fresh serum (Coppock and Pardee, 1985) (Fig. 1). The kinetic pattern of induction of TK enzyme activity correlates well with the entry of these cells into S phase as determined by labeled nuclei. TK enzyme activity continues to rise throughout S phase and is actually at its maximum in mitotic cells (J. M. Gudas, unpublished results).

The onset of DNA synthesis in BALB/c 3T3 cells appears to be regulated by a labile protein (see above). If a pulse of cycloheximide is given to serum-stimulated cells in mid-G_1, it results in an excess delay of 2 hr in the onset of DNA synthesis, and also in the induction of TK and *in vivo* TS activities (Yang and Pardee, 1986; Coppock and Pardee, 1985). Other studies have implicated IGF-1 as the growth factor that is responsible for the coordinate regulation of all three of these events at the G_1–S boundary (Yang and Pardee, 1986). Taken together these results suggest that the same cellular signal is involved in the control of all three processes. Therefore, by undertaking a detailed study of the molecular

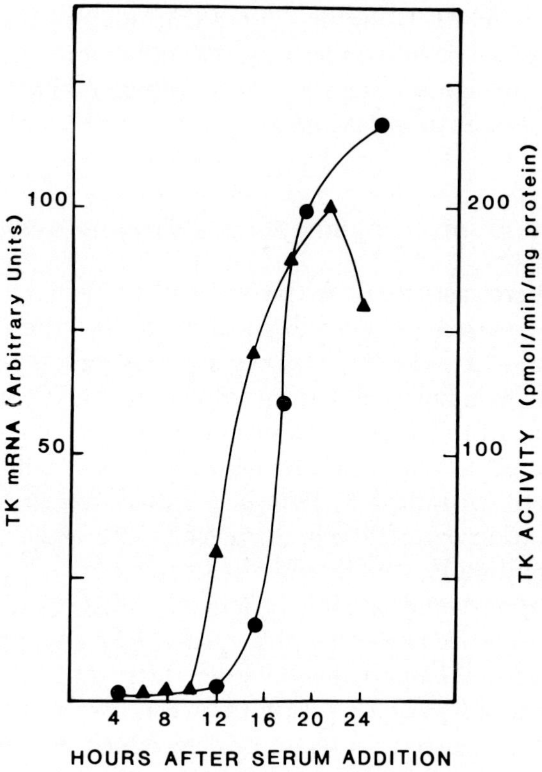

Fig. 1. Thymidine kinase enzyme activity and TK mRNA levels after serum stimulation of quiescent 3T3 cells. Cells were synchronized by serum starvation for 48 hr. Following addition of fresh serum, cells were harvested for analysis of TK enzyme activity (●) or cytoplasmic TK mRNA content (▲) at the times designated. The cytoplasmic TK mRNA data was plotted from a densitometric scan of the autoradiograph shown in Fig. 2.

mechanisms that underlie TK induction, we expect to gain a better understanding of important regulatory events that occur at the onset of DNA synthesis.

C. Regulation of Thymidine Kinase mRNA

The cytoplasmic gene for TK has been cloned from chicken (Perucho *et al.,* 1980; Merrill *et al.,* 1984b), hamster (Lewis,

1986), mouse (Lin *et al.*, 1985; Hofbauer *et al.*, 1987), and human (Bradshaw, 1983; Bradshaw and Deininger, 1984) cells, thus enabling a detailed analysis of its regulation to be undertaken in several species. We have examined the kinetics of appearance of TK mRNA in the cytoplasm of quiescent BALB/c 3T3 cells following restimulation with fresh serum. TK mRNA was first detectable in the cytoplasm at 12 hr (Fig. 2). The increase in TK mRNA preceded the subsequent burst of TK enzyme activity at 14 hr post-serum stimulation (Fig. 1). TK mRNA levels continued to rise until a peak was reached at 21 hr after serum addition. At 24 hr, when the steady-state levels of TK mRNA had declined slightly, TK enzyme activity continued to rise. The delay between the rise of TK mRNA level and the increase in TK enzyme activity at the G_1–S boundary, together with the discrepancy between TK mRNA level and enzyme activity in late S, suggest that other posttranscriptional or translational events may be involved in determining the absolute level of TK protein in the cell.

The underlying mechanism responsible for this large increase in TK mRNA at the onset of DNA synthesis has been investigated by several groups. Using nuclear run-on assays, Coppock and Pardee (1987) demonstrated a 3- to 4-fold increase in the rate of TK gene transcription at the G_1–S boundary (Fig. 3). Stewart *et al.* (1987) found a 6- to 7-fold burst in TK gene transcription at the G_1–S boundary of CV-1 cells. This increased rate of transcription declined to approximately 2- to 3-fold during S phase. The steady-state mRNA levels increase at least 20-fold during S phase in both cell types and therefore it is likely that posttranscriptional mechanisms are also involved in TK mRNA accumulation.

D. Nuclear Factor Binding to the Thymidine Kinase Promoter

Transcriptional regulation of mammalian genes appears to be mediated by specific DNA–protein interactions. This concept of transcriptional control by binding of specific activator and repressor molecules to regulatory DNA sequences was proposed 30

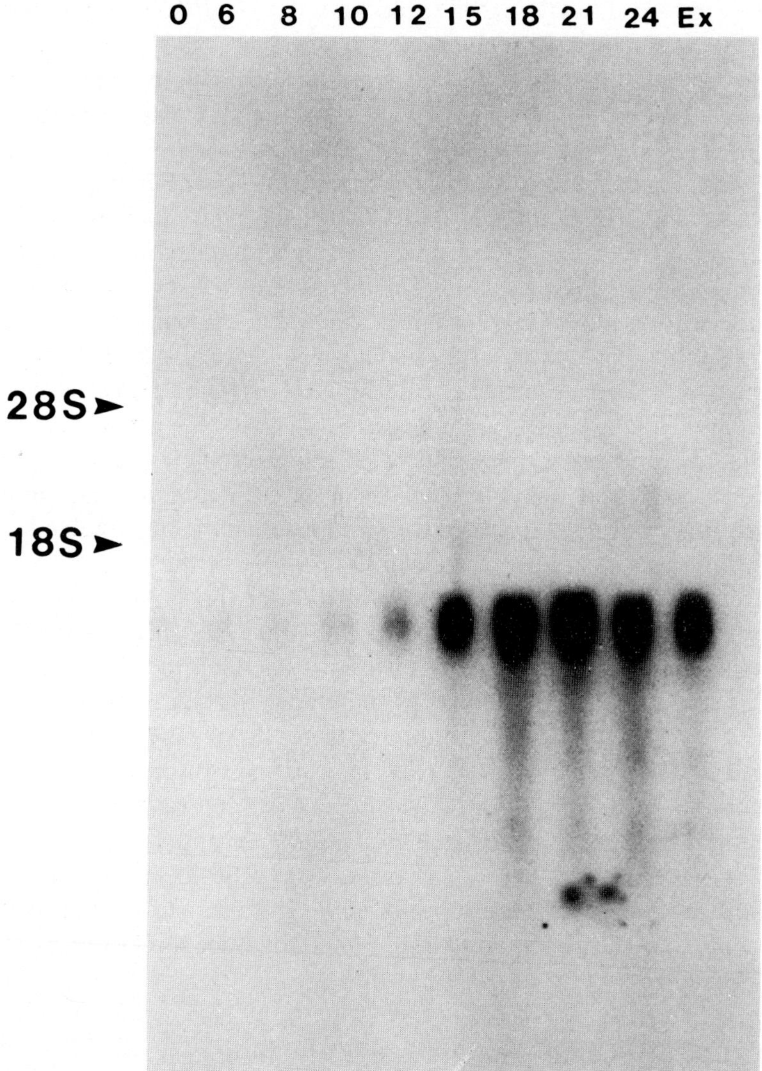

Fig. 2. Time course of cytoplasmic TK mRNA during transition of 3T3 cells from G_0 to S. Thirty μg of total RNA prepared at the designated times was run on a formaldehyde gel and transferred to Nytran. The subsequent Northern blot was hybridized to a mouse TK cDNA probe (Gudas *et al.*, 1988).

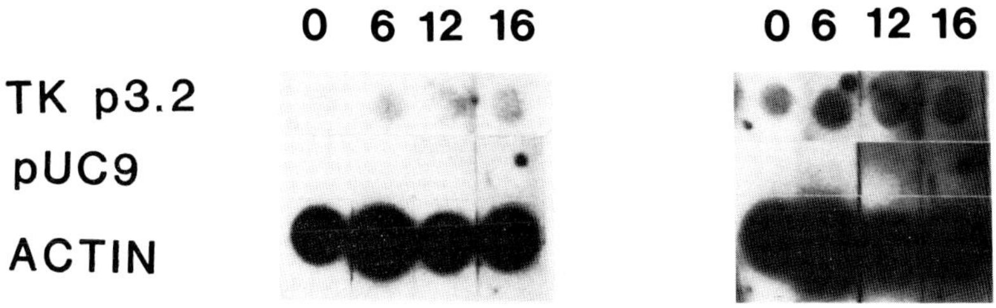

Fig. 3. Run-on transcription in isolated nuclei of 3T3 cells. Nuclear run-on transcripts labeled with [^{32}P]UTP were obtained at the times indicated after serum stimulation of quiescent cells. Equal numbers of counts were hybridized to nitrocellulose filters containing restricted and denatured plasmids TK p3.2 (10 μg), pUC9 (10 μg), and actin (1 μg). The left side is a 35-hr exposure and the right side is a 2-week exposure of the same filters (Coppock and Pardee, 1987).

years ago from studies of *lac* gene induction in *Escherichia coli* (Pardee *et al.*, 1959). The increase of TK mRNA at the G_1–S boundary is in part due to enhanced transcription of the TK gene. This may result from temporal changes in the binding of specific transcription factors to the TK promoter. To test this hypothesis, several DNA fragments spanning about 1000 bp upstream of the TK gene were used to locate regions bound by nuclear factors. Electrophoretic mobility shift assays were used to detect these interactions (Fried and Crothers, 1981; Garner and Revzin, 1981).

The TK promoter is contained within 100 bp of the transcriptional start site of the gene (Kreidberg and Kelly, 1986). Nuclear extracts derived from cells at various times during the G_0 to S phase transition formed nucleoprotein complexes with this region. Of particular significance is the finding that the nature of the complexes changed dramatically at the G_1–S boundary (Knight *et al.*, 1987) (Fig. 4). This change in the binding pattern of nucleoprotein complexes correlated well with previously reported increases in TK transcription at this time (Coppock and Pardee, 1987; Stewart *et al.*, 1987). The types of complexes found in quiescent A31 cells were retained until 9 hr after serum stimulation (complexes *a*–*c*). At 11 hr poststimulation, two new complexes (*e* and *f*), which had greater mobilities, appeared. By 12 hr

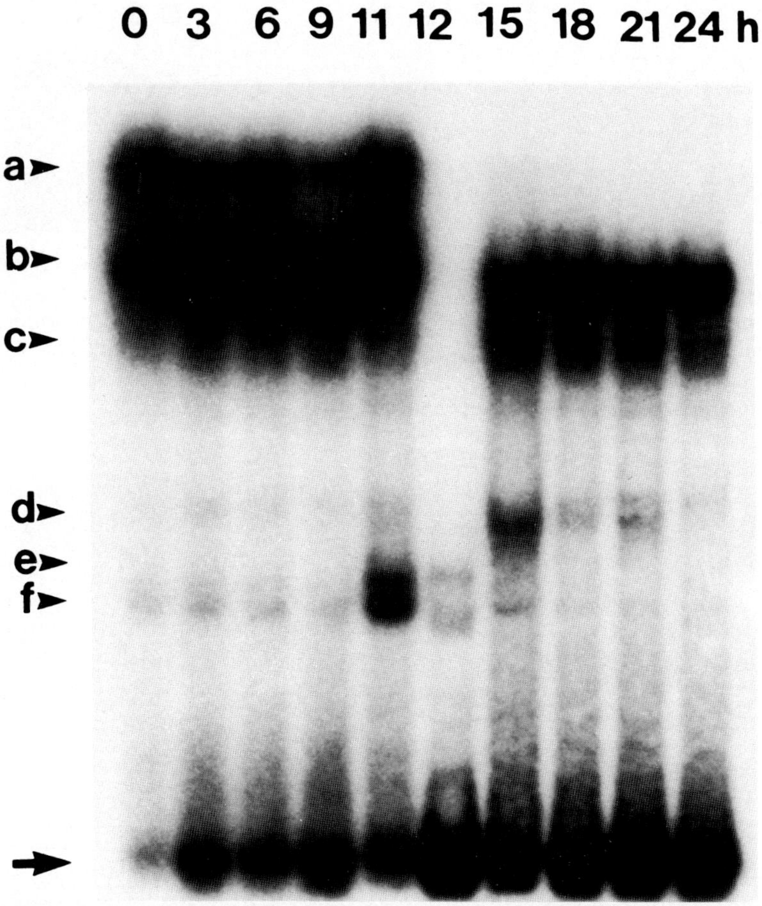

Fig. 4. Cell cycle modulation of factor binding to the TK promoter. An end-labeled 67-bp *Nco*I–*Bst*NI fragment was incubated with 12 μg of nuclear extracts prepared from cells at various times during the G_1–S transition. The time (hr) is indicated above each lane. Arrowheads indicate the nucleoprotein complexes. An arrow points to the bands for free DNA (Knight *et al.*, 1987).

(G_1–S), these were the most abundant species, owing to the apparent lack of the G_0–G_1 complexes. By 15 hr another complex (*d*) appeared, but decreased in abundance with increasing time. Also, at times later than G_1–S, the abundance of complexes *e* and *f* decreased and complexes *b* and *c* increased. Complex *a* did not re-form following the G_1–S boundary. These findings suggest that more than one protein is involved in the formation of these complexes and that dissociation of G_0–G_1 complexes results in the formation of the G_1–S complexes. Following G_1–S, the G_0–G_1 type complexes re-form by means of the putative intermediate complex *d*. The absence of complex *a* may be a function of the continued transcription of TK through S phase.

To determine the specific sequences in the human TK promoter which are bound by nuclear factors, a DNA fragment containing the sequence between −63 and +4 was used in methylation interference experiments (Fig. 5). In this technique, the nucleoprotein complexes formed with partially methylated DNA are gel-purified and the DNA is subsequently cleaved in the base-specific reaction. The sites of protein binding result in the absence or decreased intensity of specific bands in the guanosine sequence ladder (Sen and Baltimore, 1986; Weinberger *et al.*, 1986; Staudt *et al.*, 1986). Both the G_0 and G_1–S nuclear extracts give rise to the diminished intensities of two guanosines on the coding strand (Knight *et al.*, 1987) (Fig. 5A), which reside within an inverted CCAAT box at −36. Another CCAAT box found at −67 competed for protein binding. Dimethyl sulfate protection of the β-globin CCAAT box occurs at the same guanosines (Cohen *et al.*, 1986). The noncoding strand showed no binding site, probably because guanosines are absent from this region on this strand.

The fact that the inverted CCAAT sequence is specifically bound by protein in both G_0 and G_1–S nuclear extracts is consistent with the finding that the TK promoter is bound by nuclear proteins throughout the G_0–S transition. However, during this interval the composition of the nucleoprotein complexes varies. Thus, the specific CCAAT-binding factor may be one of the most important proteins in these G_1–S complexes; it may provide the scaffold upon which other proteins associate during G_0–G_1 and S.

The temporal changes in these nucleoprotein complexes suggests that the human TK gene may be transcriptionally repressed until just prior to S phase. At this time, a brief derepression may

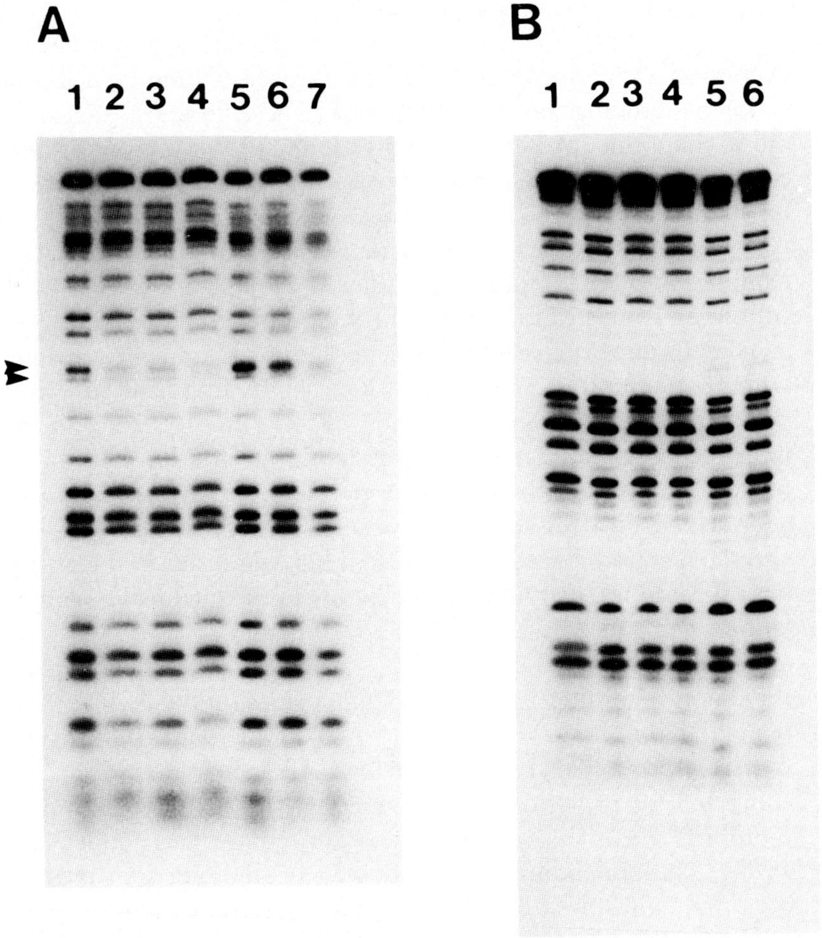

Fig. 5. Methylation interference experiments for the TK promoter. The arrowheads indicate the G residues whose methylation by DMS specifically inhibits the binding of a nuclear factor to its cognate sequence. Nucleoprotein complex and free DNA bands were excised from preparative gels and analyzed after piperidine treatment. A G ladder was also generated from the fragment alone. (A) Methylation interference of the coding strand. The 67-bp *Nco*I–*Bst*NI fragment was end-labeled at the *Bst*NI site on the coding strand. Lanes 1 and 7, G ladder. Lane 2, complex *b*. Lane 3, complex *c*. Lane 4, complexes *e* and *f*. Lanes 5 and 6, free DNA from the G_0 and the G_1–S binding reactions, respectively. (B) Methylation interference of the noncoding strand. The 67-bp *Nco*I–*Bst*NI fragment was end-labeled at the *Nco*I site on the noncoding strand. Lanes 1 to 6 are the same as in (A) (Knight *et al.*, 1987).

occur, correlating with the transient increase in TK transcription observed by Stewart *et al.* (1987). Elevated transcription of TK mRNA in S phase may also correlate with the observed presence of complex *d* and the absence of complex *a*. At later times, other processes appear to regulate the ultimate cellular abundance of TK mRNA and protein (Stewart *et al.*, 1987; Coppock and Pardee, 1987; Gudas *et al.*, 1988).

Other CCAAT-binding factors have been reported in mammalian cells. One such factor (CTF) has been purified from HeLa cell nuclear extracts, and was found to be indistinguishable from nuclear factor 1 (Jones *et al.*, 1987). Another CCAAT-binding factor (CBP) was isolated from rat liver (Graves *et al.*, 1986). Two other distinct CCAAT-binding factors (NF-Y and NF-Y*) have been described to bind the major histocompatibility complex II genes (Dorn *et al.*, 1987). Our preliminary experiments indicate the CCAAT-binding factor we observe does not correspond in size to either CTF or CBP (Y.-D. Guo and G. B. Knight, unpublished observations.

E. Posttranscriptional Regulation of Thymidine Kinase mRNA

To date, little information has been generated to further our understanding of the posttransciptional mechanisms involved in TK mRNA regulation. The TK gene was found to be transcribed in quiescent BALB/c 3T3 cells with no accumulation of mature mRNA. Because TK mRNA has a relatively long half-life (around 8 hr), the results suggest that one step of regulation may occur at the level of mRNA stability (Coppock and Pardee, 1987). Several groups have performed promoter switch experiments in which heterologous viral promoters linked to cellular TK cDNA sequences have been transfected into recipient cells and examined for cell cycle-regulated expression. These experiments suggest that genetic elements capable of conferring growth phase-dependent regulation on TK mRNA levels (Merrill *et al.*, 1984a; Stewart *et al.*, 1987; Hofbauer *et al.*, 1987) and TK enzyme activity (Lewis and Matkovich, 1986) are contained entirely within sequences of the mature cytoplasmic mRNA.

We have examined nuclear RNA levels as serum-starved cells were released from quiescence and found that a dramatic change in the processing of TK mRNA occurred at the onset of DNA synthesis. This change involved the appearance of a series of high-molecular-weight precursor bands that most likely correspond to various RNA processing intermediates. Evidence that the larger bands represent true processing intermediates derives from the determination that a TK cDNA probe hybridized to the entire series of higher molecular weight precursors while an intron probe only hybridized to a discrete subset of larger bands. Our results thus suggest that quiescent cells lack the ability to process efficiently certain mRNAs which are required during S phase, and that this function is gained just prior to the onset of DNA synthesis (Gudas *et al.*, 1988).

IV. Summary

In this chapter we have summarized evidence for the existence of regulatory points in the G_0–S phase transition of the cell cycle. Restriction point control in late G_1 involves rapid protein synthesis of a labile regulatory protein. A candidate protein subserving this role, p68, has been identified from 2-D gel analyses. The coincident occurrence of DNA synthesis and induction of TK expression, together with control of both by similar cellular signals, establishes the latter as a model for investigating G_1–S regulatory events at the molecular level. Our results show that the TK gene is subject to a variety of cellular regulatory controls, all of which act to permit the expression of TK mRNA and enzyme activity at a very specific time in the cell cycle. The discovery of a complex control mechanism, involving increased gene transcription, altered nucleoprotein binding, and more efficient processing of the TK hnRNA, suggests that an orchestrated network of regulation is activated at the G_1–S boundary. It is now our goal to dissect and study the various control mechanisms implicated in TK gene regulation. Characterization of key proteins involved in these processes and of their syntheses, interactions, and modifications, should increase our understanding of the more global processes of cell proliferation and the differences between normal and tumorigenic cells.

Acknowledgments

We would like to thank Prescott Deininger for the human λ TK46 clone and Rheinhold Hofbauer for the mouse cDNA clone. This work was supported by Public Health Service Grant GM 24571 to A.B.P. and Training Grant T32 CA09361 to G.B.K. and J.M.G.

References

Bradshaw, H. D., Jr. (1983). Molecular cloning and cell cycle-specific regulation of a functional human thymidine kinase gene. *Proc. Natl. Acad. Sci. U.S.A.* **80,** 5588–5591.

Bradshaw, H. D., Jr., and Deininger, P. L. (1984). Human thymidine kinase gene: Molecular cloning and nucleotide sequence of a cDNA expressible in mammalian cells. *Mol. Cell. Biol.* **4,** 2316–2320.

Campisi, J., and Pardee, A. B. (1984). Post-transcriptional control of the onset of DNA synthesis by an insulin-like growth factor. *Mol. Cell. Biol.* **4,** 1807–1814.

Campisi, J., Medrano, E. E., Morreo, G., and Pardee, A. B. (1982). Restriction point control of cell growth by a labile protein: Evidence for increased stability in transformed cells. *Proc. Natl. Acad. Sci. U.S.A.* **79,** 436–440.

Claycomb, W. C. (1979). DNA synthesis and DNA enzymes in terminally differentiating cardiac muscle cells. *Exp. Cell Res.* **118,** 111–114.

Cohen, R. B., Sheffery, M., and Kim, C. G. (1986). Partial purification of a nuclear protein that binds to the CCAAT box of the mouse α1-globin gene. *Mol. Cell. Biol.* **6,** 821–832.

Coppock, D. L., and Pardee, A. B. (1985). Regulation of thymidine kinase activity in the cell cycle by a labile protein. *J. Cell. Physiol.* **124,** 269–274.

Coppock, D. L., and Pardee, A. B. (1987). Control of thymidine kinase mRNA during the cell cycle. *Mol. Cell. Biol.* **7,** 2925–2932.

Croy, R. G., and Pardee, A. B. (1983). Enhanced synthesis and stabilization of M_r

68,000 protein in transformed BALB/c 3T3 cells: Candidate for restriction point control of cell growth. *Proc. Natl. Acad. Sci. U.S.A.* **80,** 4699–4703.

Dorn, A., Bollekens, J., Staub, A., Benoist, C., and Mathis, D. (1987). A multiplicity of CCAAT box-binding proteins. *Cell (Cambridge, Mass.)* **50,** 863–872.

Ferrari, S., and Baserga, R. (1987). Oncogenes and cell cycle genes. *BioEssays* **7,** 9–13.

Fried, M., and Crothers, D. M. (1981). Equilibria and kinetics of Lac repressor operator interactions by polyacrylamide gel electrophoresis. *Nucleic Acids Res.* **9,** 6505–6525.

Garner, M. M., and Revzin, A. (1981). A gel electrophoresis method for quantifying the binding of proteins to specific DNA regions: Application to components of the *E. coli* lactose operon regulatory system. *Nucleic Acids Res.* **9,** 3047–3060.

Graves, B. J., Johnson, P. F., and McKnight, S. L. (1986). Homologous recognition of a promoter domain common to the MSV LTR and the HSV *tk* gene. *Cell (Cambridge, Mass.)* **44,** 565–576.

Greenberg, M. E., and Ziff, E. B. (1984). Stimulation of 3T3 cells induces transcription of the c-*fos* proto-oncogene. *Nature (London)* **311,** 433–438.

Gudas, J. M., Knight, G. B., and Pardee, A. B. (1988). Nuclear posttranscriptional processing of thymidine kinase mRNA at the onset of DNA synthesis. *Proc. Natl. Acad. Sci. U.S.A.* **85,** 4705–4709.

Hendrickson, S., Wu, J., and Johnson, L. F. (1980). Cell cycle regulation of dihydrofolate reductase mRNA metabolism in mouse fibroblasts. *Proc. Natl. Acad. Sci. U.S.A.* **77,** 5140–5144.

Hofbauer, R., Mullner, E., Seiser, C., and Wintersberger, E. (1987). Cell cycle regulated synthesis of stable mouse thymidine kinase mRNA is mediated by a sequence within the cDNA. *Nucleic Acids Res.* **15,** 741–752.

Johnson, L. F., Fuhrman, C. L., and Weidemann, L. M. (1978). Regulation of dihydrofolate reductase gene expression in mouse fibroblasts during the transition from resting to growing state. *J. Cell. Physiol.* **97,** 397–406.

Johnson, L. F., Rao, L. G., and Muench, A. J. (1982). Regulation of thymidine kinase enzyme level in serum-stimulated mouse 3T6 fibroblasts. *Exp. Cell Res.* **138,** 79–85.

Jones, K. A., Kadonaga, J. T., Rosenfeld, P. J., Kelly, T. J., and Tjian, R. (1987). A cellular DNA-binding protein that activates eucaryotic transcription and DNA replication. *Cell (Cambridge, Mass.)* **48,** 79–89.

Kit, S. (1976). Thymidine kinase, DNA synthesis, and cancer. *Mol. Cell. Biochem.* **11,** 161–182.

Knight, G. B., Gudas, J. M., and Pardee, A. B. (1987). Cell-cycle-specific interaction of nuclear DNA-binding proteins with a CCAAT sequence from the human thymidine kinase gene. *Proc. Natl. Acad. Sci. U.S.A.* **84,** 8350–8354.

Kreidberg, J. A., and Kelly, T. J. (1986). Genetic analysis of the human thymidine kinase gene promoter. *Mol. Cell. Biol.* **6**, 2903–2909.

Kruijer, W., Cooper, J. A., Hunter, T., and Verma, I. M. (1984). Platelet-derived growth factor induces rapid but transient expression of the c-*fos* gene and protein. *Nature* (*London*) **312**, 711–716.

Leof, E. B., Wharton, W., Van Wyk, J. J., and Pledger, W. J. (1982). Epidermal growth factor (EGF) and somatomedin C regulate G_1 progression of competent BALB/c 3T3 cells. *Exp. Cell Res.* **141**, 107–115.

Lewis, J. A. (1986). Structure and expression of the Chinese hamster thymidine kinase gene. *Mol. Cell. Biol.* **6**, 1998–2010.

Lewis, J. A., and Matkovich, D. A. (1986). Genetic determinants of growth phase-dependent and adenovirus 5-responsive expression of the Chinese hamster thymidine kinase gene are contained within thymidine kinase mRNA sequences. *Mol. Cell. Biol.* **6**, 2262–2266.

Leys, E. J., and Kellems, R. (1981). Control of dihydrofolate reductase messenger ribonucleic acid production. *Mol. Cell. Biol.* **1**, 961–971.

Lin, P. F., Lieberman, H. F., Yeh, D.-B., Xu, T., Zhau, S.-Y., and Ruddle, F. H. (1985). Molecular cloning and structural analysis of murine thymidine kinase genomic and cDNA sequences. *Mol. Cell. Biol.* **5**, 3149–3156.

Merrill, G. F., Harland, R. M., Groudine, M., and McKnight, S. L. (1984a). Genetic and physical analysis of the chicken TK gene. *Mol. Cell. Biol.* **4**, 1769–1776.

Merrill, G. F., Hauschka, S. D., and McKnight, S. (1984b). TK enzyme expression in differentiating muscle cells is regulated through an internal segment of the cellular TK gene. *Mol. Cell. Biol.* **4**, 1777–1784.

Müller, R., Bravo, R., Burckhardt, J., and Curran, T. (1984). Induction of c-*fos* gene and protein by growth factors precedes activation of c-*myc*. *Nature* (*London*) **312**, 716–720.

Navalgund, L. G., Rossana, C., Muench, A. J., and Johnson, L. F. (1980). Cell cycle regulation of thymidylate synthase gene expression in cultured mouse fibroblasts. *J. Biol. Chem.* **255**, 7386–7390.

Pardee, A. B. (1987). Molecules involved in proliferation of normal and cancer cells. *Cancer Res.* **47**, 1488–1491.

Pardee, A. B., Jacob, F., and Monod, J. (1959). The genetic control and cytoplasmic expression of inducibility in the synthesis of β-galactosidase by *E. coli*. *J. Mol. Biol.* **1**, 165–178.

Perucho, M., Hanahan, D., Lipsich, L., and Wigler, M. (1980). Isolation of the chicken thymidine kinase gene by plasmid rescue. *Nature* (*London*) **285**, 207–210.

Pledger, W. J., Stiles, C. D., Antoniades, H. N., and Scher, C. D. (1977). Induction of DNA synthesis in BALB/c 3T3 cells by serum components: Re-evaluation of the commitment process. *Proc. Natl. Acad. Sci. U.S.A.* **74**, 4481–4485.

Pledger, W. J., Stiles, C. D., Antoniades, H. N., and Scher, C. D. (1978). An ordered sequence of events is required before BALB/c 3T3 cells become committed to DNA synthesis. *Proc. Natl. Acad. Sci. U.S.A.* **75,** 2839–2843.

Rossow, P., Riddle, V. G. H., and Pardee, A. B. (1979). Synthesis of labile serum dependent protein in early G_1 controls animal cell growth. *Proc. Natl. Acad. Sci. U.S.A.* **76,** 4446–4450.

Sen, R., and Baltimore, D. (1986). Multiple nuclear factors interact with the immunoglobulin enhancer sequences. *Cell (Cambridge, Mass.)* **46,** 705–716.

Staudt, L. M., Singh, H., Sen, R., Wirth, T., Sharp, P. A., and Baltimore, D. (1986). A lymphoid-specific protein binding to the octamer motif of immunoglobulin genes. *Nature (London)* **323,** 640–643.

Stewart, C. J., Ito, M., and Conrad, S. E. (1987). Evidence for transcriptional and post-transcriptional control of the cellular thymidine kinase gene. *Mol. Cell. Biol.* **7,** 1156–1163.

Stubblefield, E., and Mueller, G. C. (1965). Thymidine kinase activity in synchronized HeLa cell cultures. *Biochem. Biophys. Res. Commun.* **20,** 535–538.

Weinberger, J., Baltimore, D., and Sharp, P. A. (1986). Distinct factors bind to apparently homologous sequences in the immunoglobulin heavy-chain enhancer. *Nature (London)* **322,** 846–848.

Yang, H., and Pardee, A. B. (1986). Insulin-like growth factor I regulation of transcription and replicating enzyme induction necessary for DNA synthesis. *J. Cell. Physiol.* **127,** 410–416.

2

Negative Growth Factors in Yeast: Genetic Control of Synthesis and Response

IRA HERSKOWITZ

Department of Biochemistry and Biophysics
University of California, San Francisco
San Francisco, California

I. Introduction

Peptide growth factors regulate an important aspect of the life cycle of the yeast *Saccharomyces cerevisiae*, the mating process. Haploid cells of this yeast grow mitotically to produce progeny cells about every 100 min. When cells of opposite mating type are near each other, they mutually arrest each other's cell division cycle and begin a process that culminates in formation of a diploid cell. The two haploid mating types of yeast, **a** and α, each produce

THE REGULATION OF PROLIFERATION AND DIFFERENTIATION IN NORMAL AND NEOPLASTIC CELLS

a peptide mating pheromone, **a**-factor and α-factor, respectively. Both of these pheromones are derived from larger precursors (Brake *et al.*, cited in Gething, 1985; Kurjan and Herskowitz, 1982; Singh *et al.*, 1983; Julius *et al.*, 1983, 1984a). **a**-factor contains twelve amino acids and has an incompletely characterized modification of a Cys residue (Betz *et al.*, 1987); α-factor is thirteen amino acids in length.

The mating factors have two broad actions on target cells: They induce expression of a variety of genes whose products are necessary for steps in the mating process, and they cause cells to arrest in the cell division cycle. Formation of a diploid cell requires both cell and nuclear fusion, and mating factors activate expression of genes necessary for both of these processes (Trueheart *et al.*, 1987; McCaffrey *et al.*, 1987; Rose *et al.*, 1986). How the mating factors trigger expression of these genes (for example, the *FUS1* gene necessary for cell fusion) is not known, although DNA sequences that are responsible for this induction have been identified (Trueheart *et al.*, 1987; Kronstad *et al.*, 1987).

The mating factors cause arrest in the G_1 phase of the cell division cycle, just prior to the initiation of DNA synthesis. These mating factors inhibit progression through the cell division cycle and thus can be considered negative growth factors. The point of cell cycle arrest is the same as the point at which certain cell cycle mutants arrest, such as those defective in the *CDC28* gene (Bücking-Throm *et al.*, 1973; Hereford and Hartwell, 1974). This point within the cell cycle is regarded as a key decision-making point (*start*), at which cells make a variety of decisions, depending on the environmental circumstances: They can continue another round of cell division if nutrients are sufficient and no mating factors are present, or they can stop at this position either if nutrients are limiting or if mating factors are present. After the mating factors have acted on the mating partners, cell and nuclear fusion occurs. Because cells are arrested in G_1, each nucleus contains a full complement of chromosomes, and a diploid cell is produced. The function of the mating factors is thus to synchronize the cell cycles of the mating partners, so that a diploid is produced with exactly two copies of each chromosome. The diploid is insensitive to the mating factors and grows mitotically, giving rise to daughter diploid cells. [The diploid cell also exhibits regulation of its cell division cycle in an interesting way: When it receives the appropriate environmental stimulus, nutritional starvation, it arrests in G_1 and then undergoes meiosis. Regulation of the initiation of meiosis is discussed by Esposito and Klapholz (1981) and Mitchell and Herskowitz (1986; see also McLeod and Beach, 1988, for a description of regulation of meiosis in fission yeast by a protein kinase).]

In this chapter, I summarize our current understanding of the negative growth factors of yeast. First, I shall give a brief description of synthesis of the two different mating factors and of the machinery involved in cell response to these factors. Then I shall describe the molecular mechanisms that are responsible for the appropriate cells producing the appropriate specialized products (why **a** cells produce **a**-factor and not α-factor, etc.). Finally, I shall discuss the many different ways in which the yeast response system can be disrupted and draw some analogies with loss of growth control by mammalian cells. Other reviews on yeast mating types which focus on other aspects include Herskowitz and Oshima (1981), Sprague *et al.* (1983b), Herskowitz (1985, 1986), and Nasmyth and Shore (1987).

II. Synthesis of Mating Pheromones

There are two functional genes coding for α-factor precursors (*MFα1* and *MFα2*) (Kurjan and Herskowitz, 1982; Singh *et al.*, 1983; Kurjan, 1985). The major gene is *MFα1*, and it codes for a precursor of 165 amino acids (Kurjan and Herskowitz, 1982; Julius *et al.*, 1984a). This precursor contains four copies of the mature α-factor, each preceded by peptide processing signals of the form Lys-Arg-Glu-Ala-Glu-Ala-Glu-Ala. The first processing event is cleavage between Arg and Glu by the product of the *KEX2* gene (Julius *et al.*, 1984b). Subsequently, Glu-Ala units are removed by the product of the *STE13* gene (Julius *et al.*, 1983). Lys-Arg is removed from the carboxy terminus of α-factor by the product of the *KEX1* gene (Dmochowska *et al.*, 1987). Much less is known about the biosynthesis of **a**-factor. Once again, there are two active genes coding for an **a**-factor precursor (Brake *et al.*, cited in Gething, 1985; Michaelis and Herskowitz, 1988). The putative **a**-factor precursors coded by these genes contain 36 or 38 amino acids and include only one copy of the mature **a**-factor oligopeptide, which is preceded by putative processing sites. Three genes have been identified as being necessary for posttranslational steps in **a**-factor biosynthesis, *STE6, STE14,* and *STE16* (Rine, 1979; Blair, 1979; Wilson and Herskowitz, 1987; Kubo *et al.*, 1989). The *STE6* and *STE14* proteins are hypothesized to be involved in processing and *STE16* in localization of the precursor or in posttranslational modification. The *STE16* gene is the same as the *RAM1* gene, whose product is also needed for activity of the yeast RAS proteins (Powers *et al.*, 1986). Both the **a**-factor precursor and RAS proteins share the C-terminal sequence, Cys-aliphatic-aliphatic-X. It has been proposed that the *RAM1*, or *STE16*, protein is involved in modification of the Cys residue, which is responsible for proper cellular localization of the polypeptides (Powers *et al.*, 1986).

III. Response to Mating Pheromones

The receptor for **a**-factor is coded by the *STE3* gene, and the receptor for α-factor is coded by the *STE2* gene (Burkholder and Hartwell, 1985; Nakayama *et al.*, 1985; Hagen *et al.*, 1986). Mutants lacking these gene products are resistant to the mating factors (MacKay and Manney, 1974; Hartwell, 1980; Hagen *et al.*, 1986). Evidence that these genes are directly involved in mating factor response comes from the observations that mutants with temperature-sensitive alterations in the *STE2* gene show altered binding of α-factor *in vitro* (Jenness *et al.*, 1983) and that the *STE2* genes from *S. cerevisiae* and a related yeast, *Saccharomyces kluyveri*, are responsible for the specificity of recognition of the two corresponding α-factors (Marsh and Herskowitz, 1988). The nucleotide sequence of these genes shows that they are members of the rhodopsin–β-adrenergic receptor–muscarinic acetylcholine receptor family of integral membrane proteins. All of these proteins have seven hydrophobic segments, each of which is capable of spanning a lipid bilayer membrane. A hallmark of this class of receptor proteins is that they work via GTP-binding "coupling" proteins, which associate with the receptor on the cytoplasmic side of the plasma membrane (Stryer and Bourne, 1986). Recent work demonstrates that yeast contains these G proteins as well: Dietzel and Kurjan (1987) and Miyajima *et al.* (1987) have identified a G_α subunit (termed *SCG1* or *GPA1*); V. MacKay (personal communication) and M. Whiteway (personal communication) have identified G_β and G_α subunits (products of the *STE4* and *STE18* genes, respectively). Yeast thus utilizes a signaling system that is conserved through mammals (reviewed by Herskowitz and Marsh, 1987).

Several other genes (*STE5, STE7, STE11, STE12,* and *FAR1*) are also required for response to both **a**-factor and α-factor, but their roles are not as clear as those of *STE2, STE3, STE4,* and *STE18*. The *STE4, STE5, STE7, STE11,* and *STE12* genes are all needed for high-level transcription of the receptor genes *STE2* and *STE3* (Fields and Herskowitz, 1985; Fields *et al.*, 1988). The *STE7* and *STE11* genes have nucleotide similarity to protein kinases (Teague *et al.*, 1986). Whether these *STE* gene products act only by regulating transcription or by participating in the intracellular signaling pathway or in both processes is not known. The *FAR1* gene (F. Chang, personal communication) differs from the other genes involved in response in that its defect is solely in cell cycle arrest: Mutants defective in *FAR1* still respond to mating factors by normal induction of gene expression. The *FAR1* protein might be involved in arresting the cell cycle in G_1, or perhaps it is involved in governing the level of an intracellular signal (such that a mutation leads to sufficient signal for gene induction but not for cell cycle arrest). I shall return to the components of the signaling system later, but I would like to point out that there are at least seven different genes in addition to the receptor genes in which mutations can confer resistance to cell cycle arrest by the mating factors: *STE4, STE5, STE7, STE11, STE12, STE18,* and *FAR1*.

a cells respond only to α-factor, and α cells respond only to **a**-factor. As discussed below, the specificity of response occurs because cells synthesize only one type of receptor or the other. It has been possible to construct cells that produce the type of receptor different from their normal receptor, for example, α cells that produce the receptor for α-factor. These altered cells now respond to the ligand that corresponds to the new receptor (Bender and Sprague, 1986; Nakayama *et al.,* 1987). These observations indicate that the response pathway in **a** and α cells differs only in the type of receptor and that the intracellular machinery involved in signal transduction is the same in both cell types.

IV. Genetic Programming Responsible for Synthesis of Mating Factors and the Response System

Yeast cells are specialized in the same sense that liver cells and brain cells are specialized: They differ in the types of proteins that they synthesize. In the yeast case, there are three cell types: **a** and α cells, which are specialized for mating, and the **a**/α cell type, which is specialized for meiosis and spore formation. **a** cells produce **a**-factor and respond to α-factor; α cells produce α-factor and respond to **a**-factor; and **a**/α cells produce neither factor and respond to neither factor. A single genetic locus, the mating type locus (*MAT*) is responsible for these behaviors. Cells with the *MAT***a** allele are **a** cells, cells with the *MAT*α allele are α cells, and those with both alleles are **a**/α cells. The mating type locus is responsible for producing three regulatory activities, α1, α2, and **a**1-α2 (Strathern *et al.,* 1981). These three activities govern expression of four sets of genes whose members are scattered throughout the genome (Fig. 1).

The α-specific gene set (denoted *αsg)* is made up of genes whose transcripts are produced only in α cells (and not in **a** or in **a**/α cells). Examples of α-specific genes are *STE3* (which codes for

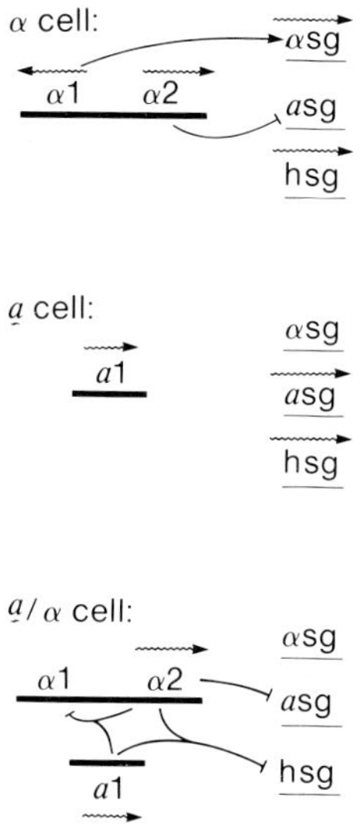

Fig. 1. Control of expression of cell-type-specific gene sets by regulatory proteins encoded by the mating type locus. The mating-type locus is shown on the left and the cell-type-specific gene sets on the right. *αsg* are α-specific genes; **a***sg* are **a**-specific genes; *hsg* are haploid-specific genes. Wavy arrows indicate transcription. Pointed arrowheads indicate stimulation of gene expression:α1 is an activator protein. Lines with blunt ends indicate inhibition of gene expression: α2 and **a**1-α2 are repressor proteins. Expression of representative α-specific genes, such as *STE3* and the two α-factor genes, is described by Sprague *et al.* (1983a) and by Fields and Herskowitz (1987). Expression of representative **a**-specific genes, such as *STE6* and the two **a**-factor genes, is described by Wilson and Herskowitz (1984) and by Michaelis and Herskowitz (1988). Expression of representative haploid-specific genes, such as *HO* and *RME,* is described by Jensen *et al.* (1983) and by Mitchell and Herskowitz (1986). Repression of the *MATα1* gene by **a**1-α2 is described by Klar *et al.* (1981), Nasmyth *et al.* (1981), Siliciano and Tatchell (1984, 1986), Miller *et al.* (1985), and by Goutte and Johnson (1988).

the receptor for **a**-factor) and the two α-factor structural genes. α1 is a DNA binding protein that recognizes a site upstream of the members of the *αsg* gene set and aids in binding of a general transcription factor (Bender and Sprague, 1987).

The **a**-specific gene set (**a***sg*) is made up of genes that are expressed only in **a** cells (and not in α or in **a**/α cells). Examples of **a**-specific genes are *STE2* (the receptor for α-factor), *STE6* (necessary for **a**-factor biosynthesis), and the two **a**-factor genes. α2 is a DNA binding protein that recognizes a site upstream of the members of the **a***sg* gene set (Johnson and Herskowitz, 1985). It may work by preventing functioning of the same general transcription factor as activates α-specific genes (Bender and Sprague, 1987; Keleher *et al.,* 1988).

The *haploid-specific* gene set (*hsg*) is made up of genes that are expressed in both **a** and α cells but not in **a**/α cells. Examples are *HO* (a site-specific endonuclease that initiates mating type switching) (Jensen *et al.,* 1983), *RME1* (which governs entry into meiosis) (Mitchell and Herskowitz, 1986), and the Ty1 element (a retrotransposon) (Elder *et al.,* 1981). Several *STE* genes (*STE4, STE5,* and *STE12*) are also turned off in **a**/α cells (see Fields and Herskowitz, 1987). The haploid-specific genes are repressed by a novel regulatory species, denoted **a**1-α2, which requires the **a**1 polypeptide coded by *MAT***a** (Kassir and Simchen, 1976) and the α2 polypeptide of *MATα*. The **a**1-α2 protein recognizes a site that is located upstream of the members of this gene set (Siliciano and Tatchell, 1984, 1986; Miller *et al.,* 1985; Goutte and Johnson, 1988).

The **a**/α-specific gene set is made up of genes that are expressed in **a**/α cells but not in **a** or α cells. Transcripts expressed in this manner have been identified (Clancy *et al.,* 1983; Percival-Smith and Segall, 1984).

α cells exhibit their characteristic cell phenotypes because the α1 protein activates transcription of α-specific genes and the α2 protein represses transcription of the **a**-specific genes. **a** cells exhibit their characteristic cell phenotypes because their **a**-specific genes are expressed (due to the absence of the repressor, α2 protein) and fail to express the α-specific genes (due to the absence of the activator, α1 protein).

a/α cells are unable to mate because they fail to express three different sets of genes that are necessary for mating: **a**-specific genes are not expressed because of the repressor, α2; α-specific genes are not expressed because **a**1-α2 turns off synthesis of their activator, α1; haploid-specific genes are also turned off by **a**1-α2. It is worth noting that **a**/α cells fail to respond to mating factors for *multiple* reasons: Not only do they not synthesize the receptor (*STE2* or *STE3* protein), but they also lack all three subunits of the G protein (*SCG1* or *GPA1, STE4, STE18*) (Dietzel and Kurjan, 1987; Miyajima *et al.*, 1987; V. MacKay and M. Whiteway, personal communication) and other products necessary for response (*STE5, STE12*) (J. Thorner, personal communication; Fields and Herskowitz, 1987).

a/α cells are able to enter meiosis because the **a**1-α2 activity represses synthesis of the product of the *RME1* gene (Mitchell and Herskowitz, 1986), which inhibits entry into meiosis.

V. Losing Response to Negative Growth Factors: Dominant Negative Mutations and Oncogenesis

Transformation of a normal cell into a cancer cell involves a genetic change that ultimately results in loss of growth control (see Varmus, 1984). *Loss of growth control* means a change in the normal requirements for growth: Cells might become constitutive for growth-stimulatory signals or hypersensitive to such signals. Another possibility is that cells become *refractory* to a growth-inhibitory signal. Our discussion will focus on the latter type of change.

We have seen that haploid yeast cells are under growth regulation by external growth factors whose natural role is coordination and facilitation of mating. As discussed above, we know a considerable amount about the way in which synthesis of the components of the response system is regulated. Because of extensive genetic analysis of the response system, we also know of at least eight different ways to become resistant to the inhibitory action of the negative growth factor. Just as yeast cells can become resistant to α-factor or **a**-factor by inactivation of any of a variety of different genes, one expects a similar situation in mammalian cells: cells that become *insensitive* to *negative* growth factors would advance at least one essential step on the road to oncogenic transformation.

Cells of yeast that respond to mating pheromones are haploid. Thus it is a simple matter to inactivate a gene and become insensitive to the mating factors. Because mammalian cells are diploid, a simple gene inactivation will not be sufficient to create a cell that is refractory to an inhibitory growth factor. Cellular transformation will result only if a cellular *deficit* is created. This can occur in at least two ways. In one case, a recessive mutation in a response gene can become homozygous or hemizygous by mitotic recombination or by chromosome loss, as occurs for retinoblastoma (Cavenee *et al.,* 1983). In another scenario, the mutational event creates an inhibitory version of the protein which behaves as an inhibitor of the wild-type protein product, in other words, a dominant negative mutation (Herskowitz, 1987).

There are several ways in which such a dominant negative mutation might be produced that would lead to loss of response to a negative growth factor. For example, let us consider a gene that is necessary to activate transcription of the receptor (or other essential component necessary for response). A mutation of its activation domain which does not disrupt its DNA binding domain may create a *trans*-acting repressor, a protein that will block action of the wild-type activator protein. Examples of this type of switching from an activator protein to a repressor protein by mutation are known (Herskowitz, 1987). In general, one expects that overproduction of such a defective product would greatly enhance its inhibitory action. Thus, one can explain why full cellular transformation by dominant negative mutations may require more than one event [as has been proposed for other types of cancer (Knudson, 1986)]. The primary event produces the inhibitory version of the product; the second event, such as gene amplification, leads to overproduction of the inhibitory product.

One can imagine alterations of other parts of the response system that could create functional deficits. The general form of these hypothetical inhibitors is that they maintain their binding ability but are defective in an effector activity. For example, consider a situation in which phosphorylation of a product is necessary for maintaining growth regulation: The phosphorylated product, for example, inhibits initiation of DNA replication by binding to an initiation protein. A mutation in this interaction domain might create a protein that, when overproduced, can tie up the protein kinase and prevent formation of the active inhibitor. Although this particular example may seem somewhat far-fetched, creation of dominant negative mutations may be more common than we realize. Given that we know some things about the components of the yeast response pathway, it might be instructive to isolate dominant, mating-factor-resistant mutants of yeast to understand the ways in which protein function can be altered to give rise to inhibitory versions.

VI. Concluding Comments

The mating types of budding yeast offer opportunities to study two major problems addressed during the conference: the molecular basis for cellular differentiation and the molecular mechanisms of growth control (and its loss). With respect to cell differentiation, we now understand at a molecular level how specialized cell types of yeast are programmed: A master regulatory locus codes for regulatory proteins that govern expression of dispersed gene sets concerned with cell specializations. With respect to growth control, we are rapidly learning how yeast cells respond to negative growth factors and the many ways in which they can become refractory to these signals by loss of gene function. Such studies reinforce the view that cellular transformation in diploid cells may sometimes result from dominant negative mutations.

Acknowledgments

I thank Fred Chang, Ken Kubo, Susan Michaelis, Vivian MacKay, and Malcolm Whiteway for communicating results prior to publication. The work from my laboratory has been supported by Research and Program Project Grants from the National Institutes of Health and from the Weingart Foundation. It is also a pleasure to acknowledge support for postdoctoral fellows from the National Institutes of Health, the Helen Hay Whitney Foundation, the Damon Runyon–Walter Winchell Cancer Fund, the Medical Research Council of Canada, and the American Cancer Society.

References

Bender, A., and Sprague, G. F., Jr. (1986). Yeast peptide pheromones **a**-factor and α-factor activate a common response mechanism in their target cells. *Cell (Cambridge, Mass.)* **47,** 929–937.

Bender, A., and Sprague, G. F., Jr. (1987). MATα1 protein, a yeast transcription activator, binds synergistically with a second protein to a set of cell-type-specific genes. *Cell (Cambridge, Mass.)* **50,** 681–691.

Betz, R., Crabb, J. W., Meyer, H. E., Wittig, R., and Duntze, W. (1987). Amino acid sequences of **a**-factor mating peptides from *Saccharomyces cerevisiae. J. Biol. Chem.* **262,** 546–548.

Blair, L. C. (1979). Genetic analysis of mating type switching in yeast. Ph.D. Thesis, University of Oregon, Eugene.

Bücking-Throm, E., Duntze, W., Hartwell, L. H., and Manney, T. R. (1973). Reversible arrest of haploid cells at the initiation of DNA synthesis by a diffusible sex factor. *Exp. Cell Res.* **76,** 99–110.

Burkholder, A. C., and Hartwell, L. H. (1985). The yeast α-factor receptor: Structural properties deduced from the sequence of the *STE2* gene. *Nucleic Acids Res.* **13,** 8463–8475.

Cavenee, W. K., Dryja, T. P., Phillips, R. A., Benedict, W. F., Godbout, R., Gallie, B. L., Murphree, A. L., Strong, L. C., and White, R. L. (1983). Expression of recessive alleles by chromosomal mechanisms in retinoblastoma. *Nature (London)* **305,** 779–784.

Clancy, M. J., Buten-Magee, B., Straight, D. J., Kennedy, A. L., Partridge, R. M., and Magee, P. T. (1983). Isolation of genes expressed preferentially during sporulation in the yeast *Saccharomyces cerevisiae*. *Proc. Natl. Acad. Sci. U.S.A.* **80**, 3000–3004.

Dietzel, C., and Kurjan, J. (1987). The yeast *SCG1* gene: A Gα-like protein implicated in the **a**- and α-factor response pathway. *Cell (Cambridge, Mass.)* **50**, 1001–1010.

Dmochowska, A., Dignard, D., Henning, D., Thomas, D. Y., and Bussey, H. (1987). Yeast *KEX1* gene encodes a putative protease with a carboxypeptidase B-like function involved in killer toxin and α-factor precursor processing. *Cell (Cambridge, Mass.)* **50**, 573–584.

Elder, R. T., St. John, T. P., Stinchcomb, D. T., and Davis, R. W. (1981). Studies on the transposable element Ty1 of yeast. I. RNA homologous to Ty1. *Cold Spring Harbor Symp. Quant. Biol.* **45**, 581–584.

Esposito, R. E., and Klapholz, S. (1981). Meiosis and ascospore development. *In* "The Molecular Biology of the Yeast Saccharomyces: Life Cycle and Inheritance" (J. N. Strathern, E. W. Jones, and J. B. Broach, eds.), pp. 211–287. Cold Spring Harbor Lab., Cold Spring Harbor, New York.

Fields, S., and Herskowitz, I. (1985). The yeast *STE12* product is required for expression of two sets of cell-type-specific genes. *Cell (Cambridge, Mass.)* **42**, 923–930.

Fields, S., and Herskowitz, I. (1987). Regulation by the yeast mating-type locus of *STE12*, a gene required for cell-type-specific expression. *Mol. Cell. Biol.* **7**, 3818–3821.

Fields, S., Chaleff, D. T., and Sprague, G. F., Jr. (1988). Yeast *STE7, STE11*, and *STE12* genes are required for expression of cell-type-specific genes. *Mol. Cell. Biol.* **8**, 551–556.

Gething, M.-J. (1985). "Protein Transport and Secretion." Cold Spring Harbor Lab., Cold Spring Harbor, New York.

Goutte, C., and Johnson, A. D. (1988). **a**1 protein alters the DNA binding specificity of α2 repressor. *Cell (Cambridge, Mass.)* **52**, 875–882.

Hagen, D. C., McCaffrey, G., and Sprague, G. F., Jr. (1986). Evidence the yeast *STE3* gene encodes a receptor for the peptide pheromone, **a** factor: Gene sequence and implications for the structure of the presumed receptor. *Proc. Natl. Acad. Sci. U.S.A.* **83**, 1418–1422.

Hartwell, L. H. (1980). Mutants of *S. cerevisiae* unresponsive to cell division control by polypeptide mating hormones. *J. Cell Biol.* **85**, 811–822.

Hereford, L. M., and Hartwell, L. H. (1974). Sequential gene function in the initiation of *Saccharomyces cerevisiae* DNA synthesis. *J. Mol. Biol.* **84**, 445–461.

Herskowitz, I. (1985). Master regulatory loci in yeast and lambda. *Cold Spring Harbor Symp. Quant. Biol.* **50**, 565–574.

Herskowitz, I. (1986). Specialized cell types in yeast: Their use in addressing problems in cell biology. *In* "Yeast Cell Biology" (J. B. Hicks, ed.), pp. 625–656. Alan R. Liss, New York.

Herskowitz, I. (1987). Functional inactivation of genes by dominant negative mutations. *Nature (London)* **329,** 219–222.

Herskowitz, I., and Marsh, L. (1987). Conservation of a receptor/signal transduction system. *Cell (Cambridge, Mass.)* **50,** 995–996.

Herskowitz, I., and Oshima, Y. (1981). Control of cell type in *Saccharomyces cerevisiae:* Mating type and mating type interconversion. *In* "The Molecular Biology of the Yeast Saccharomyces: Life Cycle and Inheritance" (J. N. Strathern, E. W. Jones, and J. R. Broach, eds.), pp. 181–209. Cold Spring Harbor Lab., Cold Spring Harbor, New York.

Jenness, D. D., Burkholder, A. C., and Hartwell, L. H. (1983). Binding of α-factor pheromone to yeast **a** cells: Chemical and genetic evidence for an α-factor receptor. *Cell (Cambridge, Mass.)* **35,** 521–529.

Jensen, R., Sprague, G. F., Jr., and Herskowitz, I. (1983). Regulation of yeast mating-type interconversion: Feedback control of *HO* gene expression by the yeast mating type locus. *Proc. Natl. Acad. Sci. U.S.A.* **80,** 3035–3039.

Johnson, A. D., and Herskowitz, I. (1985). A repressor ($MAT\alpha 2$ product) and its operator control expression of a set of cell type specific genes in yeast. *Cell (Cambridge, Mass.)* **42,** 237–247.

Julius, D., Blair, L., Brake, A., Sprague, G., and Thorner, J. (1983). Yeast α-factor is processed from a larger precursor polypeptide: The essential role of a membrane-bound dipeptidyl aminopeptidase. *Cell (Cambridge, Mass.)* **32,** 839–852.

Julius, D., Schekman, R., and Thorner, J. (1984a). Glycosylation and processing of prepro-α-factor through the yeast secretory pathway. *Cell (Cambridge, Mass.)* **36,** 309–318.

Julius, D., Brake, A., Blair, L., Kunisawa, R., and Thorner, J. (1984b). Isolation of the putative structural gene for the lysine-arginine-cleaving endopeptidase required for processing of yeast prepro-α-factor. *Cell (Cambridge, Mass.)* **37,** 1075–1089.

Kassir, Y., and Simchen, G. (1976). Regulation of mating and meiosis in yeast by the mating-type region. *Genetics* **82,** 187–206.

Keleher, C. A., Goutte, C., and Johnson, A. D. (1988). The yeast cell-type-specific repressor $\alpha 2$ acts cooperatively with a non-cell-type-specific protein. *Cell (Cambridge, Mass.)* **53,** 927–936.

Klar, A. J. S., Strathern, J. N., Broach, J. R., and Hicks, J. B. (1981). Regulation of transcription in expressed and unexpressed mating type cassettes of yeast. *Nature (London)* **289,** 239–244.

Knudson, A. G., Jr. (1986). Genetics of human cancer. *Annu. Rev. Genet.* **20,** 231–251.

Kubo, K., Michaelis, S., and Herskowitz, I. (1989). In preparation.

Kronstad, J. W., Holly, J. A., and MacKay, V. L. (1987). A yeast operator overlaps an upstream activation site. *Cell* (*Cambridge, Mass.*) **50**, 369–377.

Kurjan, J. (1985). α-factor structural gene mutations in *Saccharomyces cerevisiae:* Effects on α-factor production and mating. *Mol. Cell. Biol.* **5**, 787–796.

Kurjan, J., and Herskowitz, I. (1982). Structure of a yeast pheromone gene (*MF*α): A putative α-factor precursor contains four tandem copies of mature α-factor. *Cell* (*Cambridge, Mass.*) **30**, 933–943.

McCaffrey, G., Clay, F. J., Kelsay, K., and Sprague, G. F., Jr. (1987). Identification and regulation of a gene required for cell fusion during mating of the yeast *Saccharomyces cerevisiae. Mol. Cell. Biol.* **7**, 2680–2690.

MacKay, V., and Manney, T. R. (1974). Mutations affecting sexual conjugation and related processes in *Saccharomyces cerevisiae.* I. Isolation and phenotypic characterization of nonmating mutants. *Genetics* **76**, 255–271.

McLeod, M., and Beach, D. (1988). A specific inhibitor of the *ran1*$^+$ protein kinase regulates entry into meiosis in *Schizosaccharomyces pombe. Nature* (*London*) **332**, 509–514.

Marsh, L., and Herskowitz, I. (1988). The STE2 protein of *Saccharomyces kluyveri* is a member of the rhodopsin/beta-adrenergic receptor family and is responsible for recognition of the peptide ligand alpha factor. *Proc. Natl.* Acad. Sci. U.S.A. **85**, 3855–3859.

Michaelis, S., and Herskowitz, I. (1988). The **a**-factor pheromone of *Saccharomyces cerevisiae* is essential for mating. *Mol. Cell. Biol.* **8**, 1309–1318.

Miller, A. M., MacKay, V. L., and Masmyth, K. A. (1985). Identification and comparison of two sequence elements that confer cell-type specific transcription in yeast. *Nature* (*London*) **314**, 598–603.

Mitchell, A. P., and Herskowitz, I. (1986). Activation of meiosis and sporulation by repression of the *RME1* product in yeast. *Nature* (*London*) **319**, 738–742.

Miyajima, I., Nakafuku, M., Nakayama, N., Brenner, C., Miyajima, A., Kaibuchi, K., Arai, K., Kaziro, Y., and Matsumoto, K. (1987). *GPA1*, a haploid-specific essential gene, encodes a yeast homolog of mammalian G protein which may be involved in mating factor signal transduction. *Cell* (*Cambridge, Mass.*) **50**, 1011–1019.

Nakayama, N., Miyajima, A., and Arai, K. (1985). Nucleotide sequences of STE2 and *STE3*, cell-type-specific sterile genes from *Saccharomyces cerevisiae. EMBO J.* **4**, 2643–2648.

Nakayama, N., Miyajima, A., and Arai, K. (1987). Common signal transduction system shared by *STE2* and *STE3* in haploid cells of *Saccharomyces cerevisiae:* Autocrine cell-cycle arrest results from forced expression of *STE2. EMBO J.* **6**, 249–254.

Nasmyth, K., and Shore, D. (1987). Transcriptional regulation in the yeast life cycle. *Science* **237,** 1162–1170.

Nasmyth, K. A., Tatchell, K., Hall, B. D., Astell, C., and Smith, M. (1981). A position effect in the control of transcription at yeast mating type loci. *Nature* (*London*) **289,** 244–250.

Percival-Smith, A., and Segall, J. (1984). Isolation of DNA sequences preferentially expressed during sporulation in *Saccharomyces cerevisiae. Mol. Cell. Biol.* **4,** 142–150.

Powers, S., Michaelis, S., Broek, D., Santa Anna-A., S., Field, J., Herskowitz, I., and Wigler, M. (1986). *RAM,* a gene of yeast required for a functional modification of *RAS* proteins and for production of mating pheromone **a**-factor. *Cell* (*Cambridge, Mass.*) **47,** 413–422.

Rine, J. D. (1979). Regulation and transposition of cryptic mating type genes in *Saccharomyces cerevisiae.* Ph.D. Thesis, University of Oregon, Eugene.

Rose, M. K., Price, B. R., and Fink, G. R. (1986). *Saccharomyces cerevisiae* nuclear fusion requires prior activation by alpha factor. *Mol. Cell. Biol.* **6,** 3490–3497.

Siliciano, P. G., and Tatchell, K. (1984). Transcription and regulatory signals at the mating type locus in yeast. *Cell* (*Cambridge, Mass.*) **37,** 969–978.

Siliciano, P. G., and Tatchell, K. (1986). Identification of the DNA sequences controlling the expression of the *MATα* locus of yeast. *Proc. Natl. Acad. Sci. U.S.A.* **83,** 2320–2324.

Singh, A., Chen, E. Y., Lugovoy, J. M., Chang, C. N., Hitzeman, R. A., and Seeburg, P. H. (1983). *Saccharomyces cerevisiae* contains two discrete genes coding for the α-factor pheromone. *Nucleic Acids Res.* **11,** 4049–4063.

Sprague, G. F., Jr., Jensen, R., and Herskowitz, I. (1983a). Control of yeast cell type by the mating type locus: Positive regulation of the α-specific *STE3* gene by the *MATα1* product. *Cell* (*Cambridge, Mass.*) **32,** 409–415.

Sprague, G. F., Jr., Blair, L. C., and Thorner, J. (1983b). Cell interactions and regulation of cell type in the yeast *Saccharomyces cerevisiae. Annu. Rev. Microbiol.* **37,** 623–660.

Strathern, J., Hicks, J., and Herskowitz, I. (1981). Control of cell type in yeast by the mating type locus: The α1-α2 hypothesis. *J. Mol. Biol.* **147,** 357–372.

Stryer, L., and Bourne, H. R. (1986). G proteins: A family of signal transducers. *Annu. Rev. Cell Biol.* **2,** 391–419.

Teague, M. A., Chaleff, D. T., and Errede, B. (1986). Nucleotide sequence of the yeast regulatory gene *STE7* predicts a protein homologous to protein kinases. *Proc. Natl. Acad. Sci. U.S.A.* **83,** 7371–7375.

Trueheart, J., Boeke, J. D., and Fink, G. R. (1987). Two genes required for cell fusion during yeast conjugation: Evidence for a pheromone-induced surface protein. *Mol. Cell. Biol.* **7,** 2316–2328.

Varmus, H. E. (1984). The molecular genetics of cellular oncogenes. *Annu. Rev. Genet.* **18**, 553–612.

Wilson, K. L., and Herskowitz, I. (1984). Negative regulation of *STE6* gene expression by the $\alpha 2$ product of *Saccharomyces cerevisiae. Mol. Cell. Biol.* **4**, 2420–2427.

Wilson, K. L., and Herskowitz, I. (1987). *STE16*, a new gene for pheromone production by **a** cells of *Saccharomyces cerevisiae. Genetics* **155**, 441–449.

3

The Platelet-Derived Growth Factor: History, Chemistry, and Molecular Biology

CHARLES D. STILES

Department of Microbiology and Molecular Genetics
Harvard Medical School
and Dana-Farber Cancer Institute
Boston, Massachusetts

I. History of PDGF

Normal fibroblast cells in culture grow well if clotted blood serum is added to the culture medium. In the early 1970s, Samuel Balk and his associates made the important observation that platelet-poor plasma, the product of unclotted blood, was deficient in growth-promoting activity for fibroblast cell cultures (Balk *et al.*, 1973). They speculated that a "wound hormone" was activated or released during clot formation and that this wound hormone was required for the growth of normal fibroblasts. They observed, furthermore, that fibroblasts which had been transformed by Rous sarcoma virus had escaped the growth requirement for this wound hormone and could proliferate well in plasma-supplemented medium. In 1974, Russell Ross and his colleagues in Seattle, and at the same time Kohler and Lipton in Hershey, confirmed the original observations of Balk *et al.* and extended them by demonstrating that the platelet cell was the source of the growth factor which normal fibroblast cultures required (Ross *et al.,* 1974; Kohler and Lipton, 1974). In 1979, this platelet-derived growth factor (PDGF) was purified to homogeneity in Boston (Antoniades *et al.,* 1979) and in Uppsala (Heldin *et al.,* 1979).

The ability to work with homogeneous preparations of PDGF facilitated rapid progress in the chemistry and biology of this novel connective tissue mitogen. Primary amino acid sequence analysis of PDGF was hampered by a variety of technical problems but succeeded handsomely in 1983 with the discovery that the structural gene encoding for the B chain of PDGF (see below) was none other than the cellular homolog of the retroviral oncogene, v-*sis* (Doolittle *et al.,* 1983; Waterfield *et al.,* 1983).

Studies with radiolabeled PDGF in the early 1980s demonstrated the existence of a specific high-affinity receptor protein confined to the surface of fibroblasts, smooth muscle cells, and glial cells (Heldin *et al.,* 1981; Huang *et al.,* 1982; Glenn *et al.,* 1982). The Swedish group at Uppsala was first to demonstrate that this PDGF receptor was intimately associated with a tyrosine-specific protein kinase activity (Ek *et al.,* 1982). The tyrosine autophosphorylation response of the PDGF receptor was exploited by Rusty Williams, Axel Ullrich, and their colleagues to purify the PDGF receptor and to isolate a PDGF receptor cDNA clone. For these reasons, it came as no surprise when structural analysis of the PDGF receptor cDNA by these workers (Yarden *et al.,* 1986) revealed domains of identity with other growth factor receptors which display tyrosine kinase activity.

My own laboratory used electrophoretically homogeneous preparations of PDGF to study the molecular biology of its mitogenic action. We demonstrated in 1981 that the mitogenic response to PDGF was a transcription-dependent event (Smith and Stiles, 1981). Rare PDGF-inducible genes were isolated as cDNA clones (Cochran *et al.*, 1983). In 1983 Kathy Kelly from Philip Leder's laboratory collaborated with Brent Cochran from my own group to demonstrate that PDGF stimulates expression of the c-*myc* proto-oncogene (Kelly *et al.*, 1983). The following year, we and others (Cochran *et al.*, 1984; Greenberg and Ziff, 1984; Kruijer *et al.*, 1984; Muller *et al.*, 1984) found that c-*fos*, another oncogene encoding for a nuclear binding protein, was also PDGF-inducible.

II. Chemistry of PDGF

Native PDGF isolated from platelets consists of two separate polypeptide chains, A and B. A dimeric structure is absolutely required for PDGF mitogenic activity. Curiously, the A and B polypeptide chains of PDGF are encoded by separate genes residing on separate chromosomes in human cells. The B-chain gene, also known as c-*sis*, maps to chromosome 22 (Favera *et al.*, 1982) whereas the A-chain gene is located on chromosome 7 (Betsholtz *et al.*, 1986). Homodimers of PDGF A chain and homodimers of PDGF B chain are mitogenically active in 3T3 cell cultures, as are heterodimers consisting of an A chain and a B chain. However, the exact composition of PDGF from platelets is not well established at this time. In principle, platelet PDGF could consist entirely of A–B heterodimers, it could consist entirely of A–A and B–B homodimers, or it could comprise a mixture of all three isoforms.

III. Biology of PDGF

Although PDGF was first discovered as a platelet protein, it is now clear that there are other cellular sources of this connective tissue mitogen. Activated monocytes, together with proliferating endothelial cells, express PDGF A chain, PDGF B chain, or both together (see review by Ross *et al.*, 1986). Because PDGF is expressed or released in the context of bleeding and tissue damage and because PDGF receptors are confined to the sort of cells involved in wound healing and scar formation, many investigators in the field are comfortable with the view that PDGF functions *in vivo* to initiate and sustain the wound-healing process (as first proposed by Sam Balk). Like other growth factors, though, the biological function of PDGF may transcend those functions displayed during adult life. Growth factors are increasingly implicated as modulators of early embryonic development (reviewed by Mercola and Stiles, 1988). Our own laboratory has shown that PDGF A chain, PDGF B chain, and PDGF receptor mRNAs can be detected very early in the embryogenesis of both mice and frogs (C. D. Stiles, unpublished observations).

IV. Regulation of Gene Expression by PDGF

Our laboratory was the first to demonstrate the existence of specific, PDGF-inducible genes expressed in a way which correlated with the mitogenic response to PDGF (Cochran *et al.*, 1983). We termed these genes *competence genes* after the term coined in the mid-1970s to describe the mitogenic response of fibroblast cells to PDGF (Pledger *et al.*, 1977; Stiles *et al.*, 1979). Data from our laboratory and from others (Linzer and Nathans, 1983; Hirschhorn *et al.*, 1984; Edwards *et al.*, 1985) indicate that between 0.1 and 0.3% of the genes which are expressed in normal fibroblast cells are induced by exposure to PDGF. By extrapolation, we estimate that there are between 10 and 30 members of the PDGF-inducible competence gene family. While the c-*myc* and c-*fos* proto-oncogenes are the most visible members of the competence gene family, they are certainly not the only ones.

All of the PDGF-inducible competence genes are selectively degraded within the fibroblast cell. The c-*myc* and c-*fos* mRNAs, for example, turn over with half-lives in the range of 15–30 min whereas the average poly A^+ mRNA has a half-life of about 24 hr (Dani *et al.*, 1984; Eick *et al.*, 1985; Rabbitts *et al.*, 1985; Piechaczyk *et al.*, 1985). Selective degradation of these competence mRNAs correlates with the presence of an adenine-uridine-rich motif common to the 3′ untranslated region of these mRNAs (Caput *et al.*, 1986; Shaw and Kamen, 1986). Rapid degradation of competence mRNAs is arrested by protein synthesis inhibitors (Dani *et al.*, 1984; Linial *et al.*, 1985; Mitchell *et al.*, 1986). The ability of protein synthesis blocking agents to stabilize competence mRNAs accounts, to a large extent, for the "superinduction" of these mRNAs observed when PDGF is added together with protein synthesis inhibitors.

V. Function of Competence Genes in the Cellular Response to PDGF

Much attention has been focused on the role of c-*myc* and c-*fos* induction in the mitogenic response to PDGF and other mitogens. A full review of this literature is beyond the scope of this chapter. In brief, these studies support the view that c-*myc* and c-*fos* proteins function as intracellular mediators of the mitogenic response to PDGF; however, expression of c-*myc* and c-*fos* per se cannot totally account for the mitogenic activity of PDGF. It is likely that other PDGF-inducible gene products, together with other actions of PDGF on anabolic cell metabolism, all contribute to the mitogenic response. In addition to stimulating gene expression, for example, PDGF stimulates protein synthesis, sodium–proton exchange, phosphatidylinositol triphosphate metabolism, and arachidonic acid metabolism (see review by Rozengurt, 1986).

Less attention has been devoted to the "other" PDGF-inducible competence genes. The very first PDGF-inducible competence genes to be isolated and characterized as such were *JE* and *KC*, isolated by us in 1983 as partial length cDNA clones of moderately abundant PDGF-inducible mRNAs (Cochran *et al.*, 1983).

Very recently, we have obtained full-length cDNA clones of the

JE gene. Sequence analysis of this gene together with expression data (Rollins *et al.*, 1988) indicate that the JE protein probably does not function as an intracellular mediator of the mitogenic response. Rather, this protein is secreted from the cell. The transcription–translation product of *JE* appears to be a cytokine of some sort; however, the target cell and the biological functions of the JE protein remain to be established (B. J. Rollins and C. D. Stiles, unpublished data).

VI. Summary and Prospects

PDGF is a dimeric polypeptide which functions as a potent mitogen for fibroblasts, smooth muscle, and glial cells *in vitro*. Multiple facets of PDGF biology are consistent with the view that this molecule serves to initiate and sustain the process of wound healing *in vivo* although direct support for this hypothesis is rather thin. Other data indicate that PDGF might play a pivotal role in early embryonic development. Like all hormones, PDGF functions to regulate gene expression within its target cells. A small panel of between 10–30 PDGF-inducible competence genes has been described. Some members of the competence gene family are closely linked to genes discussed in the cancer literature. The proteins encoded by these genes, notably c-*myc* and c-*fos,* may function as intracellular mediators of the mitogenic response to PDGF. Still other PDGF-inducible genes appear to encode for cytokine-like proteins. These cytokine-like proteins may play a systemic role in the wound-healing process.

Many facets of PDGF biology remain unclear. Within the cell, the molecular events which lead to induction of competence gene expression on one hand and selective degradation of competence mRNAs on the other are largely *terra incognita*. On the other side of the plasma membrane, the mission of PDGF-inducible cytokine-like genes, such as *JE,* in the biologic response to PDGF remains to be defined.

References

Antoniades, H. N., Scher, C. D., and Stiles, C. D. (1979). Purification of human platelet-derived growth factor. *Proc. Natl. Acad. Sci. U.S.A.* **76**, 1809–1813.

Balk, S., Whitfield, J. F., Youdale, T., and Braun, A. C. (1973). Roles of calcium serum, plasma, and folic acid in the control of proliferation of normal and Rous sarcoma virus-infected chicken fibroblasts. *Proc. Natl. Acad. Sci. U.S.A.* **70**, 675–679.

Betsholtz, C., Johnson, A., Heldin, C. H., Westermark, B., Lind, P., Urdea, M. S., Eddy, R., Shows, T. B., Philpott, K., Mellot, A. L., Knott, T. J., and Scott, J. (1986). cDNA sequence and chromosomal location of human platelet-derived growth factor A chain and its expression in tumor cell lines. *Nature (London)* **320**, 697–699.

Caput, D., Beutler, B., Hartog, K., Thayer, R., Brown-Shimer, S., and Cerami, A. (1986). Identification of a common nucleotide sequence in the 3′-untranslated region of mRNA molecules specifying inflammatory mediators. *Proc. Natl. Acad. Sci. U.S.A.* **83**, 1670–1674.

Cochran, B. H., Reffel, A. C., and Stiles, C. D. (1983). Molecular cloning of gene sequences regulated by platelet-derived growth factor. *Cell (Cambridge, Mass.)* **33**, 939–947.

Cochran, B. H., Zullo, J., Verma, I. M., and Stiles, C. D. (1984). Expression of the c-*fos* gene and of a *fos*-related gene is stimulated by platelet-derived growth factor. *Science* **226**, 1080–1082.

Dani, C., Blanchard, J. M., Piechaczyk, S., El Sabouty, S., Marty, L., and Jeanteur, P. (1984). Extreme instability of *myc* mRNA in normal and transformed human cells. *Proc. Natl. Acad. Sci. U.S.A.* **81**, 7046–7050.

Doolittle, R. F., Hunkapiller, M. W., Hood, L. E., Devare, S. G., Robbins, K. C., Aaronson, S. A., and Antoniades, H. N. (1983). Simian sarcoma virus *onc* gene, v-*sis*, is derived from the gene (or genes) encoding a platelet-derived growth factor. *Science* **221**, 275–277.

Edwards, D. R., Parfett, C. L. J., and Denhardt, D. H. (1985). Transcriptional regulation of two serum-induced RNAs in mouse fibroblasts: Equivalence of one species to B2 repetitive elements. *Mol. Cell. Biol.* **5**, 3280–3288.

Eick, D., Piechaczyk, M., Henglein, B., Blanchard, J.-M., Traub, B., Kofler, E., Wiest, S., Lenoir, G. M., and Bornkamm, G. (1985). Aberrant c-*myc* RNAs of Burkitt's lymphoma cells have longer half-lives. *EMBO J.* **4**, 3717–3725.

Ek, B., Westermark, B., Wasteson, A., and Heldin, C. H. (1982). Stimulation of tyrosine-specific phosphorylation by platelet-derived growth factor. *Nature (London)* **295**, 419–420.

Favera, R. D., Gallo, R. C., Giallongo, A., and Croce, C. M. (1982). Chromosomal localization of the human homolog (c-*sis*) of the Simian Sarcoma Virus onc gene. *Science* **218,** 686–688.

Glenn, K., Bowen-Pope, D. F., and Ross, R. (1982). Platelet-derived growth factor III. Identification of a platelet-derived growth factor receptor by affinity labeling. *J. Biol. Chem.* **257,** 5172–5176.

Greenberg, M. E., and Ziff, E. B. (1984). Stimulation of 3T3 cells induces transcription of the c-*fos* proto-oncogene. *Nature* (*London*) **311,** 433–438.

Heldin, C. H., Westermark, B., and Wasteson, A. (1979). Platelet-derived growth factor: Purification and partial characterization. *Proc. Natl. Acad. Sci. U.S.A.* **76,** 3722–3726.

Heldin, C. H., Westermark, B., and Wasteson, A. (1981). Specific receptors for platelet-derived growth factor on cells derived from connective tissue and glia. *Proc. Natl. Acad. Sci. U.S.A.* **78,** 3664–3668.

Hirschhorn, R. R., Aller, P., Yuan, Z.-A., Gibson, C. W., and Baserga, R. (1984). Cell-cycle-specific cDNAs from mammalian cells temperature sensitive for growth. *Proc. Natl. Acad. Sci. U.S.A.* **81,** 6004–6008.

Huang, J. S., Huang, S. S., Kennedy, B., and Deuel, T. F. (1982). Platelet-derived growth factor. Specific binding to target cells. *J. Biol. Chem.* **257,** 8130–8136.

Kelly, K., Cochran, B. H., Stiles, C. D., and Leder, P. (1983). Specific regulation of the c-*myc* gene by lymphocyte mitogens and platelet-derived growth factor. *Cell* (*Cambridge, Mass.*) **35,** 603–610.

Kohler, N., and Lipton, A. (1974). Platelets as a source of fibroblast growth-promoting activity. *Exp. Cell Res.* **87,** 297–301.

Kruijer, W., Cooper, J. A., Hunter, T., and Verma, I. M. (1984). Platelet-derived growth factor induces rapid but transient expression of the c-*fos* gene and protein. *Nature* (*London*) **312,** 711–716.

Linial, M., Gunderson, N., and Groudine, M. (1985). Enhanced transcription of c-*myc* in bursal lymphoma cells requires continuous protein synthesis. *Science* **230,** 1126–1132.

Linzer, D. I. H., and Nathans, D. (1983). Growth-related changes in specific mRNAs of cultured mouse cells. *Proc. Natl. Acad. Sci. U.S.A.* **80,** 4271–4275.

Mercola, M., and Stiles, C. D. (1988). Growth factor superfamilies and mammalian embryogenesis. *Development* **102,** 451–460.

Mitchell, R. L., Henning-Chubb, C., Huberman, E., and Verma, I. M. (1986). C-*fos* expression is neither sufficient nor obligatory for differentiation of monomyelocytes to macrophages. *Cell* (*Cambridge, Mass.*) **45,** 497–504.

Muller, R., Bravo, R., Burckhardt, J., and Curran, T. (1984). Induction of c-*fos* gene and protein by growth factors precedes activation of c-*myc*. *Nature* (*London*) **312,** 716–720.

Piechaczyk, M., Yang, J.-Q., Blanchard, J.-M., Jeanteur, P., and Marcu, K. B.

(1985). Post-transcriptional mechanisms are responsible for accumulation of truncated c-*myc* RNAs in murine plasma cell tumors. *Cell (Cambridge, Mass.)* **42,** 589–597.

Pledger, W. J., Stiles, C. D., Antoniades, H. N., and Scher, C. D. (1977). Inductions of DNA synthesis in Balb/c-3T3 cells by serum components: Reevaluation of the commitment process. *Proc. Natl. Acad. Sci. U.S.A.* **74,** 4481–4485.

Rabbitts, P. H., Watson, J. V., Lamond, A., Forster, A., Stinson, M. A., Evan, G., Fischer, W., Atherton, E., Sheppard, R., and Rabbitts, T. H. (1985). Metabolism of c-*myc* gene products: c-*myc* mRNA and protein expression in the cell cycle. *EMBO J.* **4,** 2009–2015.

Rollins, B. J., Morrison, E. D., and Stiles, C. D. (1988). Cloning and expression of *JE,* a gene inducible by platelet-derived growth factor and whose product has cytokine-like properties. *Proc. Natl. Acad. Sci. U.S.A.* **85,** 3738–3742.

Ross, R., Glomset, J. A., Kariya, B., and Harker, L. (1974). A platelet-dependent serum factor that stimulates the proliferation of arterial smooth muscle cells *in vitro. Proc. Natl. Acad. Sci. U.S.A.* **71,** 1207–1210.

Ross, R., Raines, E. W., and Bowen-Pope, D. F. (1986). The biology of platelet-derived growth factor. *Cell (Cambridge, Mass.)* **46,** 155–169.

Rozengurt, E. (1986). Early signals in the mitogenic response. *Science* **234,** 161–166.

Shaw, G., and Kamen, R. (1986). A conserved AU sequence from the 3′ untranslated region of GM-CSF mRNA mediates selective mRNA degradation. *Cell (Cambridge, Mass.)* **46,** 658–667.

Smith, J. C., and Stiles, C. D. (1981). Cytoplasmic transfer of the mitogenic response to platelet-derived growth factor. *Proc. Natl. Acad. Sci. U.S.A.* **78,** 4363–4367.

Stiles, C. D., Capone, G. T., Scher, C. D., Antoniades, H. N., Van Wyk, J. J., and Pledger, W. J. (1979). Dual control of cell growth by somatomedins and platelet-derived growth factor. *Proc. Natl. Acad. Sci. U.S.A.* **76,** 1279–1283.

Waterfield, M. D., Scrace, G. T., Whittle, N., Stroobant, P., Johnsson, A., Wasteson, A., Westermark, B., Heldin, C. H., Huang, J. S., and Deuel, T. F. (1983). Platelet-derived growth is structurally related to the putative transforming protein $p28^{sis}$ of simian sarcoma virus. *Nature (London)* **304,** 35–39.

Yarden, Y., Escobedo, J. A., Kuang, W.-J., Yang-Feng, T. L., Daniel, T. O., Tremble, P. M., Chen, E. Y., Ando, M. E., Harkins, R. N., Franke, U., Fried, V. A., Ullrich, A., and Williams, L. T. (1986). Structure of the receptor for platelet-derived growth factor helps define a family of closely related growth factor receptors. *Nature (London)* **323,** 226–232.

PART II

Cell Proliferation and Differentiation

4

Hemopoietic Growth Factor Action

E. RICHARD STANLEY

Department of Developmental Biology and Cancer
The Albert Einstein College of Medicine
New York, New York

I. Introduction

Research efforts of the last 20 years have indicated that the production of mature blood cells from immature blood precursor cells of the hemopoietic organs (yolk sac, fetal liver, bone marrow,

THE REGULATION OF PROLIFERATION AND
DIFFERENTIATION IN NORMAL AND NEOPLASTIC CELLS

and spleen) is regulated by growth factors. Many of these growth factors are found in the circulation and/or exert their effects when administered intravenously. Thus, their pharmacological use has tremendous potential clinical application. In addition, the detailed study of the biology of these growth factors and their action is providing new information concerning the role of growth factors in development and the mechanisms underlying normal and neoplastic cell proliferation. In this chapter, I briefly summarize the state of our knowledge of the nature and action of hemopoietic growth factors in general and indicate some of the more obvious potential clinical applications. By way of illustration, I discuss in more detail the nature and action of colony-stimulating factor 1 (CSF-1), also known as macrophage colony-stimulating factor (M-CSF), one of the better studied hemopoietic growth factors.

II. Hemopoiesis

All of the mature blood and lymphoid cells are derived from a common multipotent cell which is found in the bone marrow of adult mice and humans (reviewed in Till and McCulloch, 1980). The frequency of these cells among the nucleated cells in normal adult mouse bone marrow is 10^{-5} (Harrison *et al.*, 1988). It has not been formally proved that these "stem" cells actually self-replicate in adults. However, as adults possess a finite number of these cells, they must at least have arisen by self-replication during development. In the normal steady state, they are either noncycling or cycling very slowly. They give rise to determined progenitor cells (approximately 10^{-2} of the nucleated bone marrow cells) which have a more restricted capacity for differentiation, giving rise to fewer end cell types. Some of the progenitors may be multipotent, whereas other are unipotent and only able to give rise to one end cell type. In contrast to the more primitive cells, the unipotent progenitor cell populations are actively cycling in the normal steady state as they are regulated by specific growth factors to proliferate and differentiate into mature, nondividing end cells

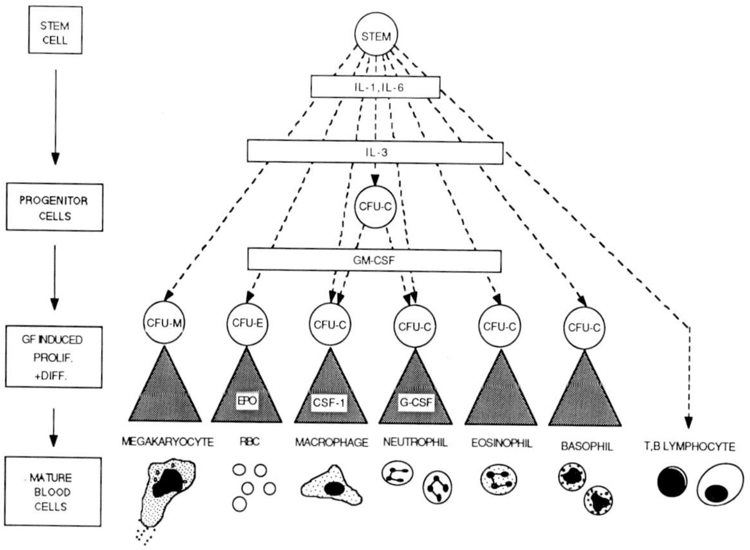

Fig. 1. Scheme of hemopoiesis indicating stages regulated by hemopoietic growth factors.

(Fig. 1). *In vitro*, the proliferative capacity of a single progenitor, although finite, can be substantial. For example, it has been calculated that one mononuclear phagocyte progenitor cell from the bone marrow can give rise to approximately 10^8 macrophage progeny (Stewart *et al.*, 1985).

III. Hemopoietic Growth Factors

Hemopoietic growth factors regulate the survival, proliferation, and differentiation of hemopoietic cells (reviewed in Stanley and Jubinsky, 1984; Metcalf, 1986; Clark and Kamen, 1987). They may be divided into two groups: those that regulate cells of a particular lineage (lineage specific) and those that regulate cells of more than one lineage (multilineage growth factors). Several of these growth factors were discovered by their ability to stimulate individual progenitor cells to form clones of morphologically recognizable blood cells and are consequently known as *colony-stimulating factors*. Others were originally described as having an immunological role and were named *interleukins*. Of particular interest is the fact that the genes for many of these growth factors [granulocyte–macrophage colony-stimulating factor (GM-CSF), interleukin-3 (IL-3), and CSF-1 (reviewed in LeBeau *et al.*, 1987)] map close together on the long arm of chromosome 5 and that $5q^-$ deletions have been associated with several disorders of the hemopoietic system, including the "$5q^-$" refractory anemia syndrome, therapy-related acute myeloid leukemia, and myeloid dysplasia (Sokal *et al.*, 1975; Kerkhofs *et al.*, 1982; Wisniewski and Hirschhorn, 1983).

A. Lineage-Specific Growth Factors

The known lineage-specific growth factors are erythropoietin (EPO), granulocyte colony-stimulating factor (G-CSF), and CSF-1. They are all found in the circulation and regulate the erythroid, neutrophilic granulocytic, and mononuclear phagocytic lineages respectively. EPO is a 34-kDa, glycosylated, single polypeptide chain (Sawyer *et al.*, 1987). Its primary site of synthesis is the tubular tissue of the kidney (Schuster *et al.*, 1987). Human EPO is encoded by a single gene of ~3.3 kb in length which maps to the long arm of chromosome 7 (7q21–7q22) and comprises five exons (Powell *et al.*, 1986; Jacobs *et al.*, 1985; Lin *et al.*, 1985). Receptors for EPO have been identified by binding studies with radiolabeled growth factor (Sawyer *et al.*, 1987; Mayeux *et al.*, 1987). G-CSF is an 18–22-kDa, glycosylated, single polypeptide chain (Nicola *et al.*, 1983). Human G-CSF is encoded by a single gene of ~3 kb in length which maps to human chromosome 17 (17q21) and comprises five exons (Nagata *et al.*, 1986a,b; Souza *et al.*, 1986; Simmers *et al.*, 1987; Tweardy *et al.*, 1987). Receptors for G-CSF have been identified on neutrophilic granulocytes by binding studies with radiolabeled growth factor (Begley *et al.*, 1988). Both CSF-1 and its receptor have been cloned and are discussed in detail in Section IV.

B. Multilineage Growth Factors

The known multilineage growth factors include those that alone regulate cells of more than one lineage (GM-CSF and IL-3) as well as others (e.g., IL-1, IL-6) which alone possess little or no hemopoietic growth factor activity. In the former group, GM-CSF is a ~22-kDa, heavily glycosylated, single polypeptide chain which stimulates the proliferation and differentiation of neutrophilic granulocytes and mononuclear phagocytes and, to some degree, erythroid progenitors and eosinophilic granulocytes. It is encoded by a single gene of ~3 kb in length which maps to human chromosome 5 (5q23–q32) and is composed of four exons (Gough *et al.*, 1984; Wong *et al.*, 1985; Miyatake *et al.*, 1985; Huebner *et al.*, 1985; LeBeau *et al.*, 1986a). Receptors for GM-CSF have been found on neutrophilic granulocytes, macrophages, and eosinophils (Walker and Burgess, 1985; DiPersio *et al.*, 1988). IL-3 is a 28-kDa, heavily glycosylated, single polypeptide chain (Ihle *et al.*, 1982) which stimulates the proliferation and differentiation of neutrophilic, eosinophilic, and basophilic granulocytes, mononuclear phagocytes, erythroid precursors, and megakaryocytic cells and their precursors. It is encoded by a single gene of ~3 kb in length which maps to human chromosome 5 (5q23–q32) within 9 kb of the GM-CSF gene, with which it shares a similar intron–exon arrangement, suggesting an evolutionary relationship (Fung *et al.*, 1984; LeBeau *et al.*, 1987; Yang *et al.*, 1988). High-affinity binding sites for IL-3 have been found on IL-3-dependent cell lines (Palaszynski and Ihle, 1984) and bone marrow cells (Nicola and Metcalf, 1986).

The other class of multilineage growth factors, which includes IL-1 and IL-6, exerts its effects on very primitive hemopoietic cells by synergism with other hemopoietic growth factors (Section III,C). There are two IL-1s, IL-1α and IL-1β, which, despite limited sequence similarity, act via the same receptor and have the same effect on hemopoietic cells. Both are ~17-kDa proteins, derived from ~31-kDa polypeptide precursors (reviewed in Lomedico *et al.*, 1986). Both IL-1α and IL-1β map to human chromosome 2 (Webb *et al.*, 1986; Dinarello, 1988). The IL-1 receptor has been shown to be an ~82-kDa glycoprotein (Urdal *et al.*, 1988). IL-6 is a glycoprotein of ~20 kDa, which has been independently identified as a 26-kDa inducible protein in fibroblasts (Haegeman *et al.*, 1986), interferon $\beta 2$ (Zilberstein *et al.*, 1986), B-cell stimulatory factor-2 (Hirano *et al.*, 1986), and as a hemopoietic growth factor (Ikebuchi *et al.*, 1987).

C. Synergism

An important breakthrough in studies of the regulation of primitive hemopoietic cell proliferation and differentiation was made by Hodgson and Bradley (1979) who demonstrated that pretreatment of mice with 5-fluorouracil led, by 48 hr after treatment, to a significant enrichment of very primitive hemopoietic cells over progenitor cells in the bone marrow. *In vitro* experiments with these populations indicated that more than one hemopoietic growth factor was required for their proliferation (Bradley and Hodgson, 1979; Kriegler *et al.*, 1982). This led to purification of the multilineage growth factor, hemopoietin-1, which possessed no hemopoietic growth factor activity alone but which allowed CSF-1 and/or IL-3 to stimulate the proliferation of more primitive hemopoietic cells than either CSF-1 or IL-3 alone or in combination (Jubinsky and Stanley, 1985; Stanley *et al.*, 1986). Hemopoietin-1 was recently shown to be IL-1α (Mochizuki *et al.*, 1987), and purified hemopoietic cell populations have been used to demonstrate that IL-1 directly stimulates primitive hemopoietic cells, thus eliminating the possibility that all of its effects were secondary and due to its stimulation of the release of other growth factors by more mature contaminating bone marrow cells (Bartelmez *et al.*, 1989). Recent studies indicate that combinations of three growth factors, IL-1, IL-3, and either IL-5 or CSF-1, provide optimal conditions for the proliferation and/or differentiation of very primitive cells to eosinophils (Warren and Moore, 1988) or macrophages (Bartelmez *et al.*, 1989), respectively. Consistent with these *in vitro* findings, *in vivo* experiments indicate that combinations of IL-1 and other hemopoietic growth factors are more effective than single growth factors in stimulating hemopoietic recovery in mice injected with cytotoxic drugs (Moore and Warren, 1987; T. R. Bradley, personal communication). IL-6 has also been shown to have an effect similar to that of IL-1. IL-6 alone may stimulate proliferation of some granulocyte or macrophage precursors, but in combination with IL-3, it enhances the rate of appearance of blast cell colonies that are supported by IL-3, apparently by reducing the G_0 period of primitive hemopoietic cells (Ikebuchi *et al.*, 1987).

IV. Colony Stimulating Factor 1

A. CSF-1 Structure and Biosynthesis

1. *Structure and Function*

CSF-1 stimulates the survival, proliferation, and differentiation of molecular phagocytes (precursor cell → monoblast → promonocyte → monocyte → macrophage) (reviewed in Stanley *et al.*, 1983). Purified murine or human CSF-1 (45–75 kDa) is a homodimer consisting of two identical ~14-kDa polypeptide chains that are maintained as a dimer by disulfide bonds (Stanley and Heard, 1977; Das and Stanley, 1982). It is heavily glycosylated, with N-linked oligosaccharides of the acidic complex type. Considerable variation in the molecular mass of the native molecule can be ascribed to variation in the degree of glycosylation. Human CSF-1 of 70–90 kDa has also been purified, suggesting that some carboxy-terminal proteolysis without loss of biological activity may have occurred in earlier preparations (Csejtey and Boosman, 1986; Wong *et al.*, 1987). Although removal of the carbohydrate does not affect its biological, antibody-binding, or receptor-binding activities, all of these activities are lost following gentle reduction of the molecule to monomeric subunits (Das and Stanley, 1982; Das *et al.*, 1981). Although studies of the nature and distribution of CSF-1-producing cells are incomplete, CSF-1 is produced by fibroblasts (Tushinski *et al.*, 1982), uterine epithelial cells (Pollard *et al.*, 1987), and stimulated blood monocytes (Horiguchi *et al.*, 1986; Rambaldi *et al.*, 1987; Oster *et al.*, 1987).

It is not produced by macrophages (Tushinski *et al.*, 1982), so its induced production by monocytes is of interest and may be relevant to the blood monocyte–tissue macrophage transition.

2. Gene Structure and Biosynthesis

Human CSF-1 is encoded by a single gene of 21 kb comprising 9 introns and 10 exons (Kawasaki *et al.*, 1985; Ladner *et al.*, 1987). The gene has been mapped to human chromosome 5q33.1, proximal to the CSF-1 receptor gene at 5q33.3 (Pettenati *et al.*, 1987; LeBeau *et al.*, 1986b). Exons 1–8 contain the coding sequence and exons 9 and 10 encode alternative 3′ untranslated regions. In human cell lines, several CSF-1 transcripts are apparent, due, at least in part, to alternative splicing and multiple alternative polyadenylation sites (Kawasaki *et al.*, 1985). In mouse tissues, cDNAs corresponding to the two major mRNA species (4.6 kb, 2.3 kb) have been sequenced (Ladner *et al.*, 1988). Apart from two single nucleotide differences, which may be polymorphisms or may have arisen due to reverse transcriptase errors, the open reading frames of the two major species are identical. The mRNAs differ in their 3′ untranslated regions, which are alternatively encoded by exons 9 and 10. Both human and murine cDNA clones predict a protein with a 32-amino acid leader sequence whose amino acid sequence extends beyond the carboxy-terminal amino acid of the mature CSF-1 subunit and includes a hydrophobic stretch of 23 amino acids (Kawasaki *et al.*, 1985; Wong *et al.*, 1987; Ladner *et al.*, 1988). CSF-1 is synthesized as a transmembrane glycoprotein which rapidly dimerizes and can subsequently appear at the cell surface with its amino-terminal portion exteriorized (Rettenmier *et al.*, 1987). Mature dimeric forms of CSF-1 are slowly released by extracellular proteolytic cleavage from this precursor. Alternatively, the growth factor is more rapidly released by a secretory mechanism involving intracellular proteolytic cleavage (Rettenmier and Roussel, 1988; K. Price and E. R. Stanley, unpublished observations). Conceivably, plasma membrane-associated CSF-1 may have important physiological roles in direct cell-to-cell regulation of receptor-bearing cells that are quite distinct from those of the secreted form.

B. The CSF-1 Receptor

1. Purification and Characterization

The CSF-1 receptor was initially identified as the site of high-affinity binding of ^{125}I-labeled CSF-1 and was shown to be selectively expressed on mononuclear phagocytes (Guilbert and Stanley, 1980; Byrne *et al.*, 1981). Chemical cross-linking studies indicated that it is a single polypeptide chain of ~165 kDa that is not covalently associated with any other protein (Morgan and Stanley, 1984). Because all of the biological effects of the growth factor, including very rapid morphological changes of membrane ruffling (at 1–2 min) and vacuole formation (within 15 min) (Tushinski *et al.*, 1982), stimulation of protein synthetic rate and inhibition of the rate of protein degradation (Tushinski and Stanley, 1983), as well as the entry of cells into S-phase (Tushinski and Stanley, 1985), are a consequence of its interaction with the receptor, the receptor was purified as a starting point for studies of CSF-1 action. The purified homogeneous receptor exhibits ligand-stimulated autophosphorylation in tyrosine as well as tyrosine kinase activity for exogeneous substrates (Yeung *et al.*, 1987).

*2. Identity with the c-*fms *Product*

Evidence from several sources indicates that the CSF-1 receptor is the c-*fms* proto-oncogene product.

1. Both molecules are tyrosine kinases of ~165 kDa (Sherr *et al.*, 1985; Yeung *et al.*, 1987).
2. Both molecules are selectively expressed in mononuclear phagocytes and in human choriocarcinoma cell lines of fetal origin (Sherr *et al.*, 1985; Rettenmier *et al.*, 1986).
3. Antiserum to the v-*fms* product selectively precipitates the receptor–CSF-1 complex from solubilized membranes but does not react with CSF-1 (Sherr *et al.*, 1985).
4. ^{125}I-labeled CSF-1 specifically binds to the gp140 v-*fms* oncogene product on cells transformed with a v-*fms*-containing retrovirus (Sacca *et al.*, 1986).

5. Cells transfected with the c-*fms* gene, but not control cells, can be stimulated to bind and/or proliferate in response to human CSF-1 (Roussel *et al.*, 1987; Rothwell and Rohrschneider, 1987).

The v-*fms* product is expressed at the cell surface with its glycosylated amino-terminal domain outside the cell and its carboxy-terminal kinase domain in the cytoplasm (Rettenmier *et al.*, 1985). While the ligand-binding domain of the v-*fms* product is not truncated (Coussens *et al.*, 1986), v-*fms* can induce growth factor independence in a CSF-1-dependent line by a nonautocrine mechanism (Wheeler *et al.*, 1986), suggesting that the v-*fms* product provides growth-stimulating signals in a ligand-independent manner. In the v-*fms* product, the carboxy-terminal 40 amino acids of the c-*fms* protein have been replaced by 11 unrelated residues (Coussens *et al.*, 1986). This truncation results in the removal of a tyrosine at position 969, which, in c-*fms*, negatively regulates cell proliferation (Roussel *et al.*, 1987). However, the truncation only potentiates transformation. Recent studies suggest that single amino acid changes within residues 1–308 in the ligand-binding domain can account for the transforming activity of the v-*fms* product (reviewed in Sherr, 1988).

3. Signal Transduction

Recent experiments with the receptors for insulin and EGF indicate that their tyrosine kinase activity is essential for signal transduction (Chen *et al.*, 1987; Honegger *et al.*, 1987; Morgan and Roth, 1987). Furthermore, experiments demonstrating that transfected genes for the CSF-1 receptor (Roussel *et al.*, 1987) or the EGF receptor (Pierce *et al.*, 1988) confer regulation, by their respective ligands, of the proliferative response of the transfected cells suggest that there is a common postreceptor mechanism for the regulation of cell proliferation in different cell types. Thus a major question concerning the action of growth factors possessing tyrosine kinase receptors is the nature and function of the intracellular physiological substrates of the receptor kinases. Preincubation of macrophage membrane preparations with CSF-1, followed by the addition of [γ-^{32}P]ATP and Mn^{2+} results in increased phosphorylation of the receptor and several other proteins on tyrosine residues. CSF-1-stimulated receptor phosphorylation is directly proportional to occupancy by CSF-1 and the stimulated incorporation is relatively stable (Jubinsky *et al.*, 1988). In similar experiments with cells preincubated with [^{32}P] phosphate, CSF-1 stimulates the tyrosine phosphorylation of at least 15 different proteins (Sengupta *et al.*, 1988) and a subset of these are detected by immunoblotting with anti-phosphotyrosine antibodies (Downing *et al.*, 1988). Most of these proteins are cytoplasmic, and differences in the order of appearance of phosphorylated proteins can be discerned, despite the fact that phosphorylation of almost all of the proteins is maximal within 30 sec of addition of ligand. Immediately following CSF-1 binding, the receptor is phosphorylated in tyrosine, followed by the phosphorylation of the other proteins in a particular order (Sengupta *et al.*, 1988). Certain autonomous mutants derived from a cell line that requires CSF-1 for growth (Morgan *et al.*, 1987) have altered patterns of protein tyrosine phosphorylation, which is consistent with a role for protein phosphorylation in the postreceptor regulation of cell proliferation (J. W. Pollard, C. J. Morgan, P. Dello Sbarba, and E. R. Stanley, unpublished observations). The proteins are possibly involved in signaling via a phosphorylation cascade and are currently being characterized. Ultimately, these very early effects of CSF-1 must be linked with the observed rapid morphological changes (Tushinski *et al.*, 1982), other rapid phenomena such as activation of the Na^+–H^+ antiporter (Vairo and Hamilton, 1988), and later events such as the stimulation of glucose uptake (Hamilton *et al.*, 1986), protein synthesis (Tushinski and Stanley, 1983), and CSF-1-induced gene expression (Bravo *et al.*, 1987; Orlofsky and Stanley, 1987).

C. Physiological Aspects of CSF-1 Action

1. *Regulation of Mononuclear Phagocyte Production and Function*

CSF-1 is found in biologically active concentrations in the circulation of mice and humans. Consistent with a humoral role for CSF-1 in regulating mononuclear phagocyte proliferation and differentiation, elevated concentrations of CSF-1 are correlated with a monocytosis in several situations, e.g., pregnancy (Bartocci *et al.,* 1986). Furthermore, intravenous injections of purified recombinant human CSF-1 increase the concentrations of blood monocytes and tissue macrophages in mice (David Hume, personal communication). CSF-1 also regulates mature, nondividing macrophages by stimulating survival (Tushinski *et al.,* 1982) as well as their production of cytokines (IL-1, interferon, CSF, and tumor necrosis factor) (Lee and Warren, 1987; Warren and Ralph, 1986; Ralph *et al.,* 1986), plasminogen activator (Hamilton *et al.,* 1980), prostaglandins (Ralph *et al.,* 1986), thromboplastin (Lyberg *et al.,* 1987), and biocidal oxygen metabolites (Wing *et al.,* 1985). Although it does not activate macrophages in the classical sense, it protects murine macrophages from lytic vesicular stomatitis virus infection (Lee and Warren, 1987), promotes tumor cell lysis, and stimulates killing of *Candida albicans* (Karbassi *et al.,* 1987; reviewed in Ralph *et al.,* 1986).

2. *Regulation of Circulating CSF-1*

Although fibroblasts have been shown to synthesize CSF-1 and this may explain the broad tissue distribution of the growth factor, the cellular source of circulating CSF-1 has not been established. Thus the mechanisms underlying regulation of CSF-1 synthesis and/or release into the blood have not been elucidated. One possible mechanism could involve proteolysis of cell surface CSF-1 at sites of local inflammation. Locally released CSF-1 could be chemotactic for macrophages as well as regulatory.

In mice, the primary physiological mechanism of clearance of CSF-1 is CSF-1 receptor-mediated endocytosis and intracellular destruction by Kupffer cells of the liver and macrophages of the spleen (Bartocci *et al.,* 1987). These macrophages probably have higher accessibility to CSF-1 because of their sinusoidal localization. This clearance mechanism provides a feedback control whereby the rate of macrophage production is determined by the number of mature macrophages. It probably represents the basic coarse control in the regulation of macrophage production. Only

5% of the circulating CSF-1 is cleared by direct filtration through the kidneys. Clearance of physiological concentrations by macrophages is rapid (half-life = 10 min) but saturable. Clearance of high concentrations is primarily by direct filtration and much slower, a desirable feature relevant to potential therapeutic use of CSF-1.

3. *CSF-1 in Pregnancy*

Early experiments indicated that high concentrations of colony-stimulating activity occur in the uterus, placenta, and fetal membranes of pregnant mice (Bradley *et al.*, 1971). Using a radioimmunoassay which detects only biologically active CSF-1, Bartocci *et al.* (1986) demonstrated that in pregnant mice the concentration of the growth factor in most tissues was only increased by ~2-fold compared with its concentration in control nonpregnant female mice. Although the circulating CSF-1 concentration was only elevated 1.4-fold, the pregnant mice exhibited a marked monocytosis. More importantly, in contrast to the small increases observed in most tissues, the uterine CSF-1 concentration steadily increased during the course of pregnancy, reaching 1000 times the normal uterine concentration by term. The placental CSF-1 concentration was the highest of the remaining tissues examined but did not increase during the course of pregnancy. Fetal CSF-1 concentration increased during the later stages of pregnancy but was comparatively low. Northern analysis of uterine mRNA using a CSF-1 probe indicated that the increase in uterine CSF-1 was due to synthesis (Pollard *et al.*, 1987). Both the 4.6-kb and 2.3-kb mRNAs found in L fibroblastoid cells (Ladner *et al.*, 1988) were found in pregnant uteri but their relative proportions in pregnant uteri (1 : 20) were the inverse of their proportions in L cells (20 : 1). This is of particular interest since, due to alternative splicing, these two mRNAs contain different 3′ untranslated regions (Section IV,A,2).

The major form in the uterus is devoid of a 3′ untranslated sequence, carried by the minor form, which confers a short mRNA half-life (Shaw and Kamen, 1986). Thus this alternative splicing event might be responsible for the accumulation of CSF-1 mRNA in the uterus. *In situ* hybridization was used to demonstrate that the luminal and glandular secretory epithelial cells of the uterus specifically synthesize CSF-1. Furthermore, synthesis by these cells is under the control of the female endocrine system, since treatments with chorionic gonadotrophin in normal female mice or a combination of progesterone and estradiol in ovariectomized mice mimic the effect of pregnancy by increasing uterine CSF-1 concentrations (Bartocci *et al.*, 1986; Pollard *et al.*, 1987). The significance of these findings becomes apparent when one considers the fact that the only other cell type reported to express the CSF-1 receptor (c-*fms* mRNA) is the trophoblast (Hoshina *et al.*, 1985). At implantation, the trophoblasts invade the endometrium and together with the decidual cells of the uterus develop into the extraembryonic tissues, including the placenta. It is likely that CSF-1 has a role in the formation and maintenance of the placenta that is separate from its regulation of mononuclear phagocyte production, since (1) the mitotic index and secretory activity of the uterine epithelium is regulated by the female sex steroids; (2) fetally derived trophoblasts enter DNA synthesis in response to CSF-1; and (3) the close proximity of the uterine secretory epithelium to placental trophoblasts would permit their local stimulation by released or cell surface CSF-1 (reviewed in Pollard *et al.*, 1987). These studies strongly suggest that endocrine-regulated uterine CSF-1 regulates placental trophoblast proliferation and differentiation; this provides a fascinating system for future studies of the role of growth factors in tissue remodeling and development.

D. CSF-1 and Its Receptor in Neoplasia

Experimental studies in animals have indicated that this growth factor and its receptor can be involved in the development of neoplasms in at least three different ways. The most obvious of these is the role of the altered form of the receptor, the v-*fms* product, in the development of feline fibrosarcomas (reviewed in

Sherr, 1988). In addition, studies by Gisselbrecht *et al.* (1987) suggest that inappropriate, early expression of a normal CSF-1 receptor on primitive hemopoietic cells may contribute to the development of myeloid leukemias in mice. Two integration sites for the replication-competent Friend murine leukemia virus (F-MuLV) in murine myeloid leukemia were cloned and one of these, utilized in approximately 20% of *in vivo* primary myeloid leukemias, was shown to be at the 5′ end of the c-*fms* (CSF-1 receptor) gene. Proviral integration in this region results in the high expression of a normal-sized c-*fms* mRNA. The third example is the inappropriate expression of the CSF-1 gene as a secondary event in the development of mononuclear phagocytic tumors *in vivo*. Intraperitoneal injections of mouse retrovirus containing the c-*myc* oncogene were shown to induce mononuclear phagocytic tumors in mice with a latency period of 8–10 weeks (Baumbach *et al.*, 1986). A series of eight tumors, monoclonal with respect to the *myc* integration site, were characterized in detail (Baumbach *et al.*, 1987). Two of these tumors were shown to synthesize and release CSF-1, and the anchorage-independent growth of one of them was shown to be substantially inhibited by a neutralizing anti-serum to purified CSF-1. Furthermore, Southern analysis of DNA from this line indicated that it had undergone a rearrangement in the CSF-1 locus. These data, together with the long lag time in the development of this tumor, are consistent with a hypothesis of the activation of the CSF-1 gene and consequent autocrine regulations by CSF-1 as a secondary event in tumor development *in vivo*. These three examples serve to illustrate the variety of ways in which the genes for this growth factor and its receptor may directly or indirectly be involved in neoplastic transformation.

E. Comparisons with Other Hemopoietic Growth Factors

CSF-1 shares many features in common with the other hemopoietic growth factors. They all stimulate the survival, proliferation, and differentiation of their target cells in a similar way. Furthermore, like GM-CSF, G-CSF, and EPO, CSF-1 is found in the circulation of normal mice (Cheers *et al.*, 1988). In addition, both GM-CSF and IL-3 have (Morla *et al.*, 1988) been shown to stimulate the rapid phosphorylation of cellular proteins in tyrosine, suggesting that, as is the case with CSF-1, tyrosine phosphorylation is involved in the transduction of their signals. In this regard, it is also of interest that GM-CSF and CSF-1 stimulate the appearance and decline of specific mRNAs with similar kinetics (Orlofsky and Stanley, 1987).

However, there already appear to be significant differences between CSF-1 and the other growth factors that override differences that might be related to their various hemopoietic target cell specificities. The CSF-1 gene structure and gene expression is significantly more complex. It is the only homodimeric hemopoietic growth factor and the only one known to exist as an intrinsic membrane protein on producing cells. Because of the structural resemblance of the CSF-1 receptor to the B-type receptor of platelet-derived growth factor (PDGF) and the fact that PDGF ligands can also be homodimeric (Heldin *et al.*, 1988), the CSF-1–CSF-1 receptor system appears to be most akin to the PDGF-B type–PDGF receptor system. Finally, CSF-1 is the only hemopoietic growth factor yet shown to regulate nonhemopoietic cells (mouse trophoblastic cells). It will therefore be of interest to study the evolution of this growth factor system in relation to the evolution of the other hemopoietic growth factor systems.

V. Potential Clinical Applications of Hemopoietic Growth Factors

Treatment of most forms of cancer involves the use of cytoxic drugs or radiation, both of which have severe effects on the hemopoietic system. Studies in experimental animals indicate that administration of hemopoietic growth factors following cytotoxic drug treatment can accelerate hemopoietic recovery (Moore and Warren, 1987; T. R. Bradley, personal communication). Should such treatment prove effective in humans, the risk of infection during the long, costly hospitalization periods may be reduced.

Furthermore, in certain cases, the availability of such treatment post-chemotherapy will enable more aggressive, and possibly more effective, chemotherapy to be used. Preliminary preclinical primate studies and studies in man have been carried out with GM-CSF with encouraging results (Donahue *et al.*, 1986; Groopman *et al.*, 1987; Brandt *et al.*, 1988). Another potential application is in bone marrow transplantation, in which case growth factor administration would be expected to accelerate hemopoietic repopulation following grafting. Eventually it may even be possible to expand primitive hemopoietic cells by *in vitro* culture with growth factors prior to their infusion into recipients. Treatment of a variety of anemias with synergistic combinations of growth factors will clearly be evaluated in the near future. The particular use of erythropoietin in correcting the anemia due to end-stage renal disease in humans has been demonstrated (Eschbach *et al.*, 1987). In addition to their use in stimulating hemopoietic reconstitution, certain growth factors, such as G-CSF, which have the ability to differentiate leukemic cells terminally, may be used to treat selected leukemic patients. Finally, administration of some of the hemopoietic growth factors might result in the generation of increased numbers of effector cells able to kill infectious agents and tumor cells.

Apart from their potential therapeutic use, these growth factors may be useful diagnostically. For example, inappropriate production of CSF-1 by tumor cells from the mononuclear phagocytic or female reproductive systems may result in high circulating concentrations of the growth factor, which could be measured by CSF-1 radioimmunoassay. In addition, screening for expression of hemopoietic growth factors and their receptors in leukemias could lead to improved classification, diagnosis, and treatment.

VI. Conclusions

The new information concerning the hemopoietic growth factors that has been generated over the last 5 years has established that they clearly have a central role in the regulation of hemopoiesis.

However, much remains to be discovered, concerning particularly the control of early hemopoiesis by the existing factors and perhaps as yet undiscovered growth and/or growth-inhibitory factors. The cloning of many of these growth factors has ensured their availability in therapeutic amounts and the prospects for their clinical use are exciting. In addition, studies of the mechanism of their action are increasing our understanding of (a) the control of cellular proliferation at the molecular level, and (b) how disruption of normal regulation leads to the neoplastic state. One of these growth factors, CSF-1, possesses a more complex gene structure and pattern of gene expression than the others and appears to have an additional role in regulating placental development.

References

Bartelmez, S. H., Bradley, T. R., Bertoncello, I., Mochizuki, D. Y., Tushinski, R. J., Stanley, E. R., Hapel, A. J., Young, I. G., Kriegler, A. B., and Hodgson, G. S. (1989). In press.

Bartocci, A., Pollard, J. W., and Stanley, E. R. (1986). Regulation of colony-stimulating factor 1 during pregnancy. *J. Exp. Med.* **164,** 956–961.

Bartocci, A., Mastrogiannis, D. S., Migliorati, G., Stockert, R. J., Wolkoff, A. W., and Stanley, E. R. (1987). Macrophages specifically regulate the concentration of their own growth factor in the circulation. *Proc. Natl. Acad. Sci. U.S.A.* **84,** 6179–6183.

Baumbach, W. R., Keath, E. J., and Cole, M. D. (1986). A mouse c-*myc* retrovirus transforms established fibroblast lines *in vitro* and induces monocyte macrophage tumors *in vivo*. *J. Virol.* **59,** 276–283.

Baumbach, W. R., Stanley, E. R., and Cole, M. D. (1987). Induction of clonal monocyte–macrophage tumors in vivo by a mouse c-myc retrovirus: Rearrangement of the CSF-1 gene as a secondary transforming event. *Mol. Cell. Biol.* **7,** 664–670.

Begley, C. G., Metcalf, D., and Nicola, N. A. (1988). Binding characteristics and proliferative action of purified granulocyte colony-stimulating factor (G-CSF) on normal and leukemic human promyelocytes. *Exp. Hematol.* **16,** 71–79.

Bradley, T. R., and Hodgson, A. S. (1979). Detection of primitive macrophage progenitor cells in mouse bone marrow. *Blood* **54,** 1446–1450.

Bradley, T. R., Stanley, E. R., and Sumner, M. A. (1971). Factors from mouse tissues stimulating colony growth of mouse bone marrow cells *in vitro*. *Aust. J. Exp. Biol. Med. Sci.* **49**, 595–603.

Brandt, S. J., Peters, W. P., Atwater, S. K., Kurtzberg, J., Borowitz, M. J., Jones, R. B., Shpall, E. J., Bast, R. C., Gilbert, C. J., and Oette, D. H. (1988). Effect of recombinant human granulocyte-marcophage colony stimulating factor on hematopoietic reconstitution after high-dose chemotherapy and autologous bone marrow transplantation. *N. Engl. J. Med.* **318**, 869–876.

Bravo, R., Neuberg, M., Burckhardt, J., Almendral, J., Willich, R., and Muller, R. (1987). Involvement of common and cell type-specific pathways in c-*fos* gene control: Stable induction by cAMP in macrophages. *Cell (Cambridge, Mass.)* **48**, 251–260.

Byrne, P. V., Guilbert, L. J., and Stanley, E. R. (1981). The distribution of cells bearing receptors for a colony stimulating factor (CSF-1) in murine tissues. *J. Cell Biol.* **91**, 848–853.

Cheers, C., Haigh, A. M., Kelso, A., Metcalf, D., Stanley, E. R., and Young, A. M. (1988). Production of colony stimulating factors during infection with an intracellular bacterium: Separate determination of M-CSF, G-CSF, GM-CSF, and Multi-CSF. *Infect. Immun.* **56**, 247–251.

Chen, W. S., Lazar, C. S., Poenie, M., Tsien, R. Y., Gill, G. N., and Rosenfeld, M. A. (1987). Requirement for intrinsic protein tyrosine kinase in the immediate and late actions of the EGF receptor. *Nature (London)* **328**, 820–823.

Clark, S. C., and Kamen, R. (1987). The human hematopoietic colony-stimulating factors. *Science* **236**, 1229–1237.

Coussens, L., Van Beveren, C., Smith, D., Chen, E., Mitchell, R. L., Isacke, C. M., Verma, I. M., and Ullrich, A. (1986). Structural alteration of viral homologue of receptor proto-oncogene *fms* at carboxy terminus. *Nature (London)* **320**, 227–280.

Csejtey, J., and Boosman, A. (1986). Purification of human macrophage colony stimulating factor (CSF-1) from medium conditioned by pancreatic carcinoma cells. *Biochem. Biophys. Res. Commun.* **138**, 238–245.

Das, S. K., and Stanley, E. R. (1982). Structure–function studies of a colony stimulating factor (CSF-1). *J. Biol. Chem.* **257**, 13679–13684.

Das, S. K., Stanley, E. R., Guilbert, L. J., and Forman, L. W. (1981). Human colony stimulating factor (CSF-1) radioimmunoassay: Resolution of three subclasses of human colony stimulating factors. *Blood* **58**, 630–641.

Dinarello, C. A. (1988). Biology of interleukin-1. *FASEB J.* **2**, 108–115.

DiPersio, J., Billing, P., Kaufman, S., Eghtesady, P., Williams, R. E., and Gasson, J. C. (1988). Characterization of the human granulocyte-macrophage colony stimulating factor receptor. *J. Biol. Chem.* **263**, 1834–1841.

Donahue, R. E., Wang, E. A., Stone, D. K., Kamen, R., Wong, A. A., Sehgal,

P. K., Nathan, D. G., and Clark, S. C. (1986). Stimulation of haematopoiesis in primates by continuous infusion of recombinant GM-CSF. *Nature* (*London*) **321,** 872–875.

Downing, J. R., Rettenmier, C. W., and Sherr, C. J. (1988). Ligand-induced tyrosine kinase activity of the colony-stimulating factor 1 receptor in a murine macrophage cell line. *Mol. Cell. Biol.* **8,** 1795–1799.

Eschbach, J. W., Egrie, J. C., Downing, M. R., Browne, J. K., and Adamson, J. W. (1987). Correction of the anemia of end-stage renal disease with recombinant human erythropoietin: Results of a phase I and II clinical trial. *N. Engl. J. Med.* **316,** 73–78.

Fung, M. C., Hapel, A. J., Ymer, S., Cohen, D. R., Johnson, R. M., Campbell, H. D., and Young, I. G. (1984). Molecular cloning of cDNA for murine interleukin-3. *Nature* (*London*) **307,** 233–237.

Gisselbrecht, S., Fichelson, S., Sola, B., Bordereaux, D., Hampe, A., Andre, C., Galibert, F., and Tambourin, P. (1987). Frequent c-*fms* activation by proviral-insertion in mouse myeloblastic leukaemias. *Nature* (*London*) **329,** 259–261.

Gough, N. M., Gough, J., Metcalf, D., Kelso, A., Grail, D., Nicola, N. A., Burgess, A. W., and Dunn, A. R. (1984). Molecular cloning of cDNA encoding a murine haematopoietic growth regulator, granulocyte-macrophage colony stimulating factor. *Nature* (*London*) **309,** 763–767.

Groopman, J. E., Mitsuyasu, R. T., Deleo, M. J., Oette, D. H., and Golde, D. W. (1987). Effect of recombinant human granulocyte-macrophage colony stimulating factor on myelopoiesis in the acquired immunodeficiency syndrome. *N. Engl. J. Med.* **317,** 593–598.

Guilbert, L. J., and Stanley, E. R. (1980). Specific interaction of murine colony-stimulating factor with mononuclear phagocytic cells. *J. Cell Biol.* **85,** 153–159.

Haegeman, G., Content, J., Volckaert, G., Derynck, R., Tavernier, J., and Fiers, W. (1986). Structural analysis of the sequence coding for an inducible 26-kDa protein in human fibroblasts. *Eur. J. Biochem.* **159,** 625–632.

Hamilton, J. A., Stanley, E. R., Burgess, A. W., and Shadduck, R. K. (1980). Stimulating of macrophage plasminogen activator production by colony stimulating factors. *J. Cell. Physiol.* **103,** 435–445.

Hamilton, J. A., Vairo, G., and Lingelbach, S. R. (1986). CSF-1 stimulates glucose uptake in murine bone marrow-derived macrophages. *Biochem. Biophys. Res. Commun.* **138,** 445–454.

Harrison, D. E., Astle, C. M., and Lerner, C. (1988). Number and continuous proliferative pattern of transplanted primitive immunohematopoietic stem cells. *Proc. Natl. Acad. Sci. U.S.A.* **85,** 822–826.

Heldin, C. H., Backstrom, A., Ostman, A., Hammacher, A., Ronnstrand, L., Rubin, K., Nister, M., and Westermark, B. (1988). Binding of different dimeric forms of PDGF to human fibroblasts: Evidence for two separate receptor types. *EMBO J.* **7,** 1387–1390.

Hirano, T., Yasukawa, K., Harada, H., Taga, T., Watanabe, Y., Matsuda, T., Kashiwamura, S., Nakajima, K., Koyama, K., Iwamatsu, A., Tsunasawa, S., Sakiyama, F., Matsui, H., Takahara, Y., Taniguchi, T., and Kishimoto, T. (1986). Complementary DNA for a novel human interleukin (BSF-2) that induces B lymphocytes to produce immunoglobulin. *Nature (London)* **324,** 73–77.

Hodgson, G. S., and Bradley, T. R. (1979). Properties of haematopoietic stem cells surviving 5-fluorouracil treatment: Evidence for a pre-CFU-S cell? *Nature (London)* **281,** 381–382.

Honegger, A. M., Szapary, D., Schmidt, A., Lyall, R., Van Obberghen, E., Dull, T. J., Ullrich, A., and Schlessinger, J. (1987). Point mutation at the ATP binding site of EGF receptor abolishes protein-tyrosine kinase activity and alters cellular routing. *Cell (Cambridge, Mass.)* **51,** 199–209.

Horiguchi, J., Warren, M. K., Ralph, P., and Kufe, D. (1986). Expression of the macrophage specific colony-stimulating factor (CSF-1) during human monocytic differentiation. *Biochem. Biophys. Res. Commun. et al.,* **141,** 924–930.

Hoshina, M., Hishio, A., Bo, M., Boime, I., and Mochizuki, M. (1985). The expression of oncogene fms in human chorionic tissue. *Acta. Obstet. Gynecol. Jpn.* **37,** 2791–2798.

Huebner, K., Isobe, M., Croce, C. M., Golde, D. W., Kaufman, S. E., and Gasson, J. C. (1985). The human gene encoding GM-CSF is at 5q21-q32, the chromosome region deleted in the 5q− anomaly. *Science* **230,** 1282–1285.

Ihle, J. N., Keller, J., Henderson, L., Klein, F., and Palaszynski, E. (1982). Procedures for the purification of interleukin-3 to homogeneity. *J. Immunol. et al.,* **129,** 2431–2435.

Ikebuchi, K., Wong, G. G., Clark, S. C., Ihle, J. N., Hirai, Y., and Ogawa, M. (1987). Interleukin 6 enhancement of interleukin 3-dependent proliferation of multipotential hemopoietic progenitors. *Proc. Natl. Acad. Sci. U.S.A.* **84,** 9035–9039.

Jacobs, K., Shoemaker, C., Rudersdorf, R., Neill, S. D., Kaufman, R. J., Mufson, A., Seehra, J., Jones, S. S., Hewick, R., Fritsch, E. F., Kawakita, M., Shimizu, T., and Miyake, T. (1985). Isolation and characterization of genomic and cDNA clones of human erythropoietin. *Nature (London)* **313,** 806–810.

Jubinsky, P. T., and Stanley, E. R. (1985). Purification of hemopoietin-1, a multilineage hemopoietic growth factor. *Proc. Natl. Acad. Sci. U.S.A.* **82,** 2764–2768.

Jubinsky, P. T., Yeung, Y. G., Sacca, R., Li, W., and Stanley, E. R. (1988). Colony stimulating factor-1 stimulated macrophage membrane protein phosphorylation. *In* "Biology of Growth Factors: Molecular Biology, Oncogenes, Signal Transduction and Clinical Applications" (J. E. Kudlow, D. H. MacLennan, A. Bernstein, and A. I. Gotlieb, eds.), pp. 75–90. Plenum, New York.

Karbassi, A., Becker, J. M., Foster, J. S., and Moore, R. N. (1987). Enhanced killing of *Candida albicans* by murine macrophages treated with macrophage colony-stimulating factor: Evidence for augmented expression of mannose receptors. *J. Immunol.* **139,** 417–421.

Kawasaki, E. S., Ladner, M. B., Wang, A. M., Van Arsdell, J. N., Warren, M. K., Coyne, M. Y., Stanley, E. R., Ralph, P., and Mark, D. F. (1985). Molecular cloning of a complementary DNA encoding human macrophage-specific colony-stimulating factor (CSF-1). *Science* **230,** 291–296.

Kerkhofs, H., Hagemeijer, A., Leeksma, C. H. W., Abels, J., Den Ottolander, G. J., Somers, R., Gerrits, W. B. J., Langen Luiyen, M. M. A. C., von dem Borne, A. E. G., Jr., Van Hemel, J. O., and Geraedts, J. P. M. (1982). The 5q− chromosome abnormality in haematological disorders: A collaborative study of 34 cases from the Netherlands. *Br. J. Haematol.* **52,** 365–381.

Kriegler, A. B., Bradley, T. R., Januszewicz, E., Hodgson, G. S., and Elms, E. F. (1982). Partial purification and characterization of a growth factor for macrophage progenitor cells with high proliferative potential in mouse bone marrow. *Blood* **60,** 503–508.

Ladner, M. B., Martin, G. A., Noble, J. A., Wikoloff, D. M., Tal, R., Kawasaki, E. S., and White, T. J. (1987). Human CSF-1: Gene structure and alternative splicing of mRNA precursors. *EMBO J.* **6,** 2693–2698.

Ladner, M. B., Martin, G. A., Van Arsdell, J., Whittman, V., Warren, K., McGrogan, M., Shadle, P., and Stanley, E. R. (1988). cDNA cloning and expression of murine CSF-1 from L929 cells. *Proc. Natl. Acad. Sci. U.S.A.* **85,** 6706–6710.

LeBeau, M. M., Pettenati, M. J., Lemons, R. S., Diaz, M. O., Westbrook, C. A., Larson, R. A., Sherr, C. J., and Rowley, J. D. (1986a). Assignment of the GM-CSF, CSF-1, and FMS genes to human chromosome 5 provides evidence for linkage of a family of genes regulating hematopoiesis and for their involvement in the deletion (5q) in myeloid disorders. *Cold Spring Harbor Symp. Quant. Biol.* **51,** 899–909.

LeBeau, M. M., Westbrook, C. A., Diaz, M. O., Larsen, R. A., Rowley, J. D., Gasson, J. C., Golde, D. W., and Sherr, C. J. (1986b). Evidence for the involvement of GM-CSF and FMS in the deletion (5q) in myeloid disorders. *Science* **231,** 984–989.

LeBeau, M. M., Epstein, N. D., O'Brien, S. J., Nienhuis, A. W., Yang, Y. C., Clark, S. C., and Rowley, J. D. (1987). The interleukin 3 gene is located on human chromosome 5 and is deleted in myeloid leukemias with a deletion of 5q. *Proc. Natl. Acad. Sci. U.S.A.* **84,** 5913–5917.

Lee, M. T., and Warren, M. K. (1987). CSF-1-induced resistance to viral infection in murine macrophages. *J. Immunol.* **138,** 3019–3022.

Lin, F. K., Suggs, S., Lin, C.-H., Browne, J. K., Smalling, R., Egrie, J. C., Chen, K. K., Fox, G. M., Martin, F., Stabinsky, Z., Badrawi, S. M., Lai, P.-H., and Goldwasser, E. (1985). Cloning and expression of the human erythropoietin gene. *Proc. Natl. Acad. Sci. U.S.A.* **82,** 7580–7584.

Lomedico, P. T., Kilian, P. L., Gubler, U., Stern, A. S., and Chizzonite, R. (1986). Molecular biology of interleukin-1. *Cold Spring Harbor Symp. Quant. Biol.* **51,** 631–639.

Lyberg, T., Stanley, E. R., and Prydz, H. (1987). Colony-stimulating factor-1 induces thromboplastin activity in murine macrophages and human monocytes. *J. Cell. Physiol.* **132,** 367–370.

Mayeux, P., Billat, C., and Jacquot, R. (1987). The erythropoietin receptor of rat erythroid progenitor cells. Characterization and affinity cross-linkage. *J. Biol. Chem.* **262,** 13985–13990.

Metcalf, D. (1986). The molecular biology and functions of the granulocyte-macrophage colony-stimulating factors. *Blood* **67,** 257–267.

Miyatake, S., Otsuka, T., Yokota, T., Lee, F., and Arai, K. (1985). Structure of the chromosomal gene for granulocyte-macrophage colony stimulating factor: Comparison of the mouse and human genes. *EMBO J.* **4,** 2561–2568.

Mochizuki, D. Y., Eisenman, J. R., Conlon, P. J., Larson, A. D., and Tushinski, R. J. (1987). Interleukin-1 regulates hematopoietic activity, a role previously ascribed to hemopoietin 1. *Proc. Natl. Acad. Sci. U.S.A.* **84,** 5267–5271.

Morgan, C. J., and Stanley, E. R. (1984). Chemical crosslinking of the mononuclear phagocyte specific growth factor CSF-1 to its receptor at the cell surface. *Biochem. Biophys. Res. Comm.* **119,** 35–41.

Morgan, C. J., Pollard, J. W., and Stanley, E. R. (1987). Isolation and characterization of a cloned growth factor dependent macrophage cell line, BAC1.2F5. *J. Cell. Physiol.* **130,** 420–427.

Morgan, D. O., and Roth, R. A. (1987). Acute insulin action requires insulin receptor kinase activity: Introduction of an inhibitory monoclonal antibody into mammalian cells blocks the rapid effects of insulin. *Proc. Natl. Acad. Sci. U.S.A.* **84,** 41–45.

Morla, A. O., Schreurs, J., Miyajima, A., and Wang, J. Y. J. (1988). Hematopoietic growth factors activate the tyrosine phosphorylation of distinct sets of proteins in interleukin-3-dependent murine cell lines. *Mol. Cell. Biol.* **8,** 2214–2218.

Moore, M. A. S., and Warren, D. J. (1987). Interleukin-1 and G-CSF synergisms: *In vivo* stimulation of stem cell recovery and hematopoietic regeneration following 5-fluorouracil treatment of mice. *Proc. Natl. Acad. Sci. U.S.A.* **84,** 7134–7138.

Nagata, S., Tsuchiya, M., Asano, S., Kaziro, Y., Yamazaki, T., Yamamoto, O.,

Hirata, Y., Kubota, N., Oheda, M., Nomura, H., and Ono, M. (1986a). Molecular cloning and expression of cDNA for human granulocyte colony-stimulating factor. *Nature* (*London*) **319**, 415–418.

Nagata, S., Tsuchiya, M., Asano, S., Yamamoto, O., Hirata, Y., Kubota, N., Oheda, M., Nomura, H., and Yamazaki, T. (1986b). The chromosomal gene structure and two mRNA's for human granulocyte colony-stimulating factor. *EMBO J.* **5**, 575–581.

Nicola, N. A., and Metcalf, D. (1986). Binding of iodinated multipotential colony-stimulating factor (interleukin-3) to murine bone marrow cells. *J. Cell. Physiol.* **128**, 180–188.

Nicola, N. A., Metcalf, D., Matsumoto, M., and Johnson, G. R. (1983). Purification of a factor inducing differentiation in murine myelomonocytic leukemia cells. Identification as granulocyte colony-stimulating factor. *J. Biol. Chem.* **258**, 9017–9023.

Orlofsky, A., and Stanley, E. R. (1987). CSF-1 induced gene expression in macrophages: Dissociation from the mitogenic response. *EMBO J.* **6**, 2947–2952.

Oster, W., Lindeman, A., Horn, S., Mertelsmann, R., and Herrmann, F. (1987). Tumor Necrosis Factor (TNF)-alpha but not TNF-beta induces secretion of colony stimulating factor for macrophages (CSF-1) by human monocytes. *Blood* **70**, 1700–1703.

Palaszynski, E., and Ihle, J. M. (1984). Evidence for specific receptors for interleukin 3 on lymphokine-dependent cell lines established from long-term bone marrow cultures. *J. Immunol.* **132**, 1872–1878.

Pettenati, M. J., LeBeau, M. M., Lemons, R. S., Shima, E. A., Kawasaki, E. S., Larson, R. A., Sherr, C. J., Diaz, M. O., and Rowley, J. D. (1987). Assignment of CSF-1 to 5q 33.1: Evidence for clustering of genes regulating hematopoiesis and for their involvement in the deletion of the long arm of chromosome 5 in myeloid disorders. *Proc. Natl. Acad. Sci. U.S.A.* **84**, 2970–2974.

Pierce, J. H., Ruggiero, M., Fleming, T. P., DiFiore, P. P., Greenberger, J. S., Varticovski, L., Schlessinger, J., Rovera, G., and Aaronson, S. A. (1988). Signal transduction through the EGF receptor transfected in IL-3-dependent hematopoietic cells. *Science* **239**, 628–631.

Pollard, J. W., Bartocci, A., Arceci, R., Orlofsky, A., Ladner, M. B., and Stanley, E. R. (1987). Apparent role of the macrophage growth factor, CSF-1, in placental development. *Nature* (*London*) **330**, 484–486.

Powell, J. S., Berkner, K. L., Lebo, R. V., and Adamson, J. W. (1986). Human erythropoietin gene: High level expression in stably transfected mammalian cells and chromosomal localization. *Proc. Natl. Acad. Sci. U.S.A.* **83**, 6465–6469.

Ralph, P., Warren, M. K., Ladner, M. D., Kawasaki, E. S., Boosman, A., and

White, T. J. (1986). Molecular and biological properties of human macrophage growth factor, CSF-1. *Cold Spring Harbor. Symp. Quant. Biol.* **51**, 679–683.

Rambaldi, A., Young, D. C., and Griffen, J. D. (1987). Expression of the M-CSF (CSF-1) gene by human monocytes. *Blood* **69**, 1409–1413.

Rettenmier, C. W., and Roussel, M. F. (1988). Differential processing of colony-stimulating factor-1 precursors encoded by two human cDNAs. *Mol. Cell. Biol.* **8**, 5026–5034.

Rettenmier, C. W., Roussel, M. F., Quinn, C. O., Kitchingman, G. R., Look, A. T., and Sherr, C. J. (1985). Transmembrane orientation of glycoproteins encoded by the v-*fms* oncogene. *Cell (Cambridge, Mass.)* **40**, 971–981.

Rettenmier, C. W., Sacca, R., Furman, W. L., Roussel, M. F., Holt, J., Nienhuis, A. W., Stanley, E. R., and Sherr, C. J. (1986). Expression of the human c-*fms* proto-oncogene product (CSF-1 receptor) on peripheral blood mononuclear cells and choriocarcinoma cell lines. *J. Clin. Invest.* **77**, 1740–1746.

Rettenmier, C. W., Roussel, M. F., Ashmun, R. A., Ralph, P., Price, K., and Sherr, C.J. (1987). Synthesis of membrane-bound CSF-1 and down modulation of CSF-1 receptors in NIH-3T3 cells transformed by cotransfection of the human CSF-1 and c-*fms* (CSF-1 receptor) genes. *Mol. Cell. Biol.* **7**, 2378–2387.

Rothwell, V. M., and Rohrschneider, L. R. (1987). Murine c-*fms* cDNA: Cloning, sequence, analysis and retroviral expression. *Oncogene Res.* **1**, 311–324.

Roussel, M. F., Dull, T. J., Rettenmier, C. W., Ralph, P., Ullrich, A., and Sherr, C. J. (1987). Transforming potential of the c-*fms* proto-oncogene (CSF-1 receptor). *Nature (London)* **325**, 549–552.

Sacca, R., Stanley, E. R., Sherr, C. J., and Rettenmier, C. W. (1986). Specific binding of the mononuclear phagocyte colony stimulating factor, CSF-1, to the product of the v-*fms* oncogene. *Proc. Natl. Acad. Sci. U.S.A.* **83**, 3331–3335.

Sawyer, S. T., Krantz, S. B., and Goldwasser, E. (1987). Binding and receptor-mediated endocytosis of erythropoietin in Friend virus-infected erythroid cells. *J. Biol. Chem.* **262**, 5554–5562.

Schuster, S. J., Wilson, J. H., Erslev, A. J., and Caro, J. (1987). Physiologic regulation and tissue localization of renal erythropoietin messenger RNA. *Blood* **70**, 316–318.

Sengupta, A., Liu, W.-K., Yeung, D. C. Y., Frackelton, A. R., Jr., and Stanley, E. R. (1988). Identification and subcellular localization of proteins that are rapidly phosphorylated in tyrosine in response to CSF-1. *Proc. Natl. Acad. Sci. U.S.A.* **85**, 8062–8066.

Shaw, G., and Kamen, R. (1986). A conserved AU sequence from the 3′

untranslated region of GM-CSF mRNA mediates selective mRNA degradation. *Cell (Cambridge, Mass.)* **46**, 659–667.

Sherr, C. J. (1988). The *fms* oncogene. *Biochim. Biophys. Acta* **948(2)**, 225–243.

Sherr, C. J., Rettenmier, C. W., Sacca, R., Roussel, M. F., Look, A. T., and Stanley, E. R. (1985). The c-*fms* proto-oncogene product is related to the receptor for the mononuclear phagocyte growth factor, CSF-1. *Cell (Cambridge, Mass.)* **41**, 665–676.

Simmers, R. N., Webber, L. M., Shannon, M. F., Garson, O. M., Wong, A., Vadas, M. A., and Sutherland, G. R. (1987). Localization of the G-CSF gene on chromosome 17 proximal to the breakpoint in the t (15;17) in acute promyelocytic leukemia. *Blood* **70**, 330–333.

Sokal, G., Michaux, J. L., Van Den Berghe, H., Cordier, A., Rodhain, J., Ferrant, A., Moriau, M., De Bruyere, M., and Sonnet, J. (1975). A new hematologic syndrome with a distinct karyotype: The 5q− chromosome. *Blood* **46**, 519–533.

Souza, L. M., Boone, T. C., Gabrilove, J., Lai, P. H., Zsebo, K. M., Murdock, D. C., Chazin, V. R., Bruszewski, J., Lu, H., Chen, K. K., Barendt, J., Platzer, E., Moore, M. A. S., Mertelsmann, R., and Welte, K. (1986). Recombinant human granulocyte colony-stimulating factor: Effects on normal and leukemic myeloid cells. *Science* **232**, 61–65.

Stanley, E. R., and Heard, P. M. (1977). Factors regulating macrophage production and growth. Purification and some properties of the colony stimulating factor from medium conditioned by mouse L cells. *J. Biol. Chem.* **252**, 4305–4312.

Stanley, E. R., and Jubinsky, P. T. (1984). Factors affecting growth and differentiation of haemopoietic cells in culture. *Clin. Haematol.* **13**, 329–336.

Stanley, E. R., Guilbert, L. J., Tushinski, R. J., and Bartelmez, S. H. (1983). CSF-1—A mononuclear phagocyte lineage-specific hemopoietic growth factor. *J. Cell Biochem.* **21**, 151–159.

Stanley, E. R., Bartocci, A., Patinkin, D., Rosendaal, M., and Bradley, T. R. (1986). Regulation of very primitive multipotent, hemopoietic cells by hemopoietin-1. *Cell (Cambridge, Mass.)* **45**, 667–674.

Stewart, C. C., Walker, E. B., Johnson, C., and Little, R. (1985). Clonal analysis of bone marrow and macrophage cultures. *In* "Mononuclear Phagocytes: Characteristics, Physiology and Function" (R. van Furth, ed.), pp. 255–266. Martinus Nijhoff Publishers, Dordrecht, The Netherlands.

Till, J. E., and McCulloch, E. A. (1980). Hemopoietic stem cell differentiation. *Biochim. Biophys. Acta* **605**, 431–459.

Tushinski, R. J., and Stanley, E. R. (1983). The regulation of macrophage protein turnover by a colony stimulating factor (CSF-1). *J. Cell. Physiol.* **116**, 67–75.

Tushinski, R. J., and Stanley, E. R. (1985). The regulation of mononuclear phagocyte entry into S phase by the colony stimulating factor, CSF-1. *J. Cell. Physiol.* **122,** 221–228.

Tushinski, R. J., Oliver, I. T., Guilbert, L. J., Tynan, P. W., Warner, J. R., and Stanley, E. R. (1982). Survival of mononuclear phagocytes depends on a lineage-specific growth factor that the differentiated cells selectively destroy. *Cell (Cambridge, Mass.)* **28,** 71–81.

Tweardy, D. J., Cannizzaro, L. A., Palumbo, A. P., Shane, S., Huebner, K., Vantuiene P., Ledbetter, D. H., Finan, J. B., Nowell, P. C., and Rovera, G. (1987). Molecular cloning and characterization of a cDNA for human granulocyte colony-stimulating factor (G-CSF) from a glioblastoma multiform cell line and localization of the G-CSF gene to chromosome band 17q21. *Oncogene Res.* **1,** 209–220.

Urdal, D. L., Call, S. M., Jackson, J. L., and Dower, S. K. (1988). Affinity purification and chemical analysis of the interleukin-1 receptor. *J. Biol. Chem.* **263,** 2870–2877.

Vairo, G., and Hamilton, J. A. (1988). Activation and proliferation signals in murine macrophages. Stimulation of Na^+, K^+-ATPase activity by hemopoietic growth factors and other agents. *J. Cell. Physiol.* **134,** 13–24.

Walker, F., and Burgess, A. W. (1985). Specific binding of radioiodinated granulocyte-macrophage colony-stimulating factor to hemopoietic cells. *EMBO J.* **4,** 933–940.

Warren, D. J., and Moore, M. A. S. (1988). Synergism among interleukin 1, interleukin 3, and interleukin 5 in the production of eosinophils from primitive hemopoietic stem cells. *J. Immunol.* **140,** 94–99.

Warren, M. K., and Ralph, P. (1986). Macrophage growth factor CSF-1 stimulates human monocyte production of interferon, tumor necrosis factor and colony-stimulating activity. *J. Immunol.* **137,** 2281–2285.

Webb, A. C., Collins, K. L., Auron, P. E., Eddy, R. L., Nakai, H., Byers, M. G., Haley, L. L., Henry, W. M., and Shows, T. B. (1986). Interleukin gene (IL-1) assigned to long arm of human chromosome 2. *Lymphokine Res.* **5,** 77–85.

Wheeler, E. G., Rettenmier, C. W., Look, A. T., and Sherr, C. J. (1986). The v-*fms* oncogene induces factor independence and tumorigenicity in CSF-1 dependent macrophage cell line. *Nature (London)* **324,** 377–380.

Wing, E. J., Ampel, N. M., Waheed, A., and Shadduck, R. K. (1985). Macrophage colony-stimulating factor (M-CSF) enhances the capacity of murine macrophages to secrete oxygen reduction products. *J. Immunol.* **135,** 2052–2060.

Wisniewski, L. P., and Hirschhorn, K. (1983). Acquired partial deletions of the long arm of chromosome 5 in hematologic disorders. *Am. J. Hematol.* **15,** 295–310.

Wong, G. G., Witek, J. S., Temple, P. A., Wilkens, K. M., Leary, A. C., Luxenberg, D. P., Jones, S. S., Brown, E. L., Kay, R. M., Orr, E. C., Shoemaker, C., Golde, D. W., Kaufman, R. J., Hewick, R. M., Wang, E. A., and Clark, S. C. (1985). Human GM-CSF: Molecular cloning of the complementary DNA and purification of the natural and recombinant proteins. *Science* **228**, 810–815.

Wong, G. G., Temple, P. A., Leary, A. C., Witek-Giannotti, J. S., Yang, Y.-C., Ciarletta, A. B., Chung, M., Murtha, P., Kriz, R., Kaufman, R. J., Ferenz, C. R., Sibley, B. S., Turner, K. J., Hewick, R. M., Clark, S. C., Yanai, N., Yokota, H., Yamada, M., Saito, M., Motoyoshi, K., and Takaku, F. (1987). Human CSF-1: Molecular cloning and expression of 4-kb cDNA encoding the human urinary protein. *Science* **235**, 1504–1508.

Yang, Y.-C., Kovacic, S., Kriz, R., Wolf, S., Clark, S. C., Wellems, T. E., Nienhuis, A., and Epstein, N. (1988). The human genes for GM-CSF and IL-3 are closely linked in tandem on chromosome 5. *Blood* **71**, 958–961.

Yeung, Y. G., Jubinsky, P. T., Sengupta, A., Yeung, C. D. Y., and Stanley, E. R. (1987). Purification of the colony-stimulating factor 1 receptor and demonstration of its tyrosine kinase activity. *Proc. Natl. Acad. Sci. U.S.A.* **84**, 1268–1271.

Zilberstein, A., Ruggieri, R., Korn, J. H., and Revel, M. (1986). Structure and expression of small cDNA and genes for human interferon-beta-2, a distinct species inducible by growth stimulatory cytokines. *EMBO J.* **5**, 2529–2537.

5

Lymphocyte Activation: Role of Cell Adhesion Molecules

P. ANDERSON, C. MORIMOTO, AND S. F. SCHLOSSMAN
Division of Tumor Immunology
Dana-Farber Cancer Institute
Boston, Massachusetts

I. Introduction

It has become increasingly clear that the aberrant expression of one or more of a panel of genes involved in cellular proliferation can contribute directly to the process of malignant transformation (*1*). The overproduction or mutation of soluble growth factors or their receptors can, under experimental conditions, result in uncontrolled proliferation characteristic of the transformed phenotype (*2*). An understanding of cellular transformation is therefore predicated on an understanding of normal processes of cell growth. The activation of T lymphocytes proceeds through an ordered cell cycle progression that is characteristic of all dividing cells. Unlike many cell types, lymphocyte activation is triggered initially through cell–cell interactions mediated by intercellular adhesion molecules. Lymphocytes and their hemopoietic accessory cells express a panel of structurally related molecules capable of mediating cell adhesion (*3*, *4*). Lymphocyte-specific structures such as CD2 (T11), CD4 (T4), and CD8 (T8) bind specifically to the accessory cell surface molecules LFA-3, class II MHC, and class I MHC, respectively. Such interactions differ from soluble ligand–receptor associations in that only molecules present in the area of cell contact can participate. Furthermore, the particulate nature of the interacting cells affords a support which tends to aggregate those ligands involved in cell–cell adhesion. As a consequence of these constraints, the repertoire of adhesion molecules expressed on antigen-presenting cells will dictate a group of lymphocyte surface molecules that will coaggregate in regions of cell–cell contact (*4*). We have examined the possibility that the coaggregation of specific lymphocyte surface structures might contribute to the regulation of T cell activation.

II. Coaggregation of CD3–T-Cell Receptor with Individual Accessory Molecules Modulates T-Cell Activation

Resting T cells were cultured with Sepharose-immobilized antibodies for 4 days in the presence or absence of interleukin-2 (IL-2) (*5*, *6*). Lymphocyte proliferation was quantitated using [^{3}H]thymidine incorporation during a 16-hr pulse. As shown in Fig. 1, Sepharose-immobilized antibody failed to induce the proliferation of accessory cell-depleted T cells in the absence of IL-2. In the presence of recombinant IL-2 (5 U/ml), proliferation was observed in the presence of immobilized anti-CD3, but not in the presence of immobilized anti-CD4 or anti-CD8. In the presence of IL-2, Sepharose beads conjugated to both anti-CD3 and anti-CD4 induced a proliferative response which was significantly greater than that induced by an equal density of anti-CD3 alone (stimulation index = 3.2 ± 1.1 fold, n = 7). The combination of Sepharose (CD3) and Sepharose (CD4) failed to enhance proliferation in a

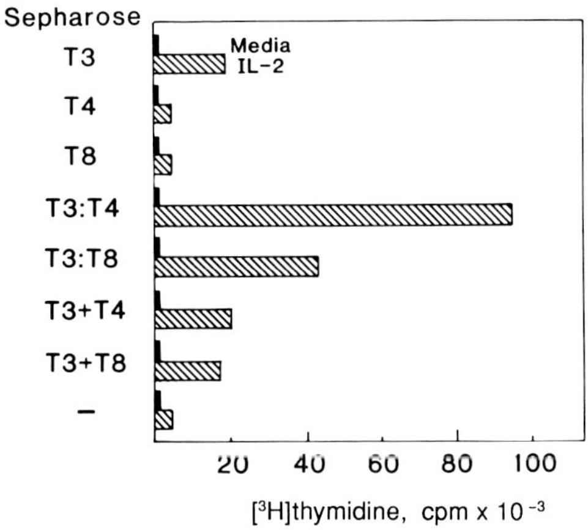

Fig. 1. Proliferative responses of T cells to immobilized antibodies: T lymphocytes were cultured in triplicate samples in round-bottom 96-well microtiter plates (Intermed, Denmark) containing 4×10^4 cells per well in RMPI containing 10% human serum (Pel-Freeze) at 37°C in a humidified CO_2 incubator. Sepharose beads were added at a concentration of 1 ml packed beads per well. When combinations of immobilized antibodies were used, each was added at a concentration of 0.5 ml packed beads per well. Where indicated, human recombinant IL-2 (Roche Labs, Nutley, NJ) was added at a concentration of 5 U/ml. Cultures were continued for 4 days at which point they were pulsed with [^{3}H]thymidine (0.2 mCi/well) (New England Nuclear, Boston, MA) for 16 hr. Cells were then harvested using a Mash II apparatus, and [^{3}H]thymidine incorporation was measured using a Packard liquid scintillation counter. Reported results represent the means of triplicate determinations in which standard errors were 15%.

similar manner. Sepharose (CD3–CD8) also induced more proliferation than Sepharose (CD3) or Sepharose (CD3) plus Sepharose (CD8). These results suggest that cross-linking of CD3 with either CD4 or CD8 enhances the proliferation induced by immobilized anti-CD3 alone. This effect was not observed following costimulation using identical antibody combinations immobilized on separate beads.

The specificity of the proliferative enhancement produced by Sepharose (CD3 : CD4) was tested using purified T-cell subpopulations. As shown in Fig. 2, Sepharose (CD3–CD4) enhanced the proliferation of CD4+ cells over that produced by Sepharose

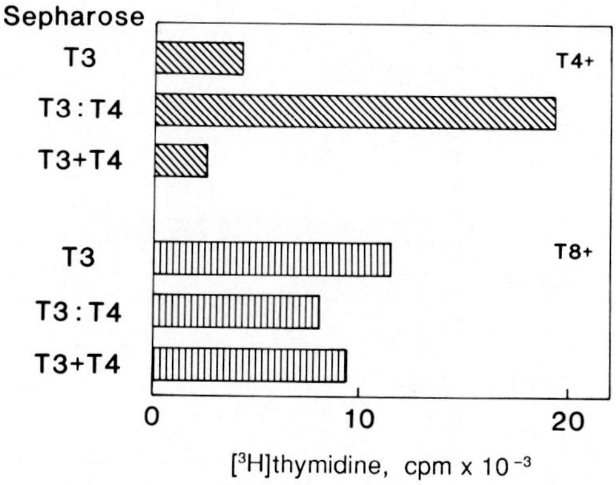

Fig. 2. Proliferative response of T cell subpopulations to immobilized antibodies: Designated populations of T cells were stimulated by immobilized antibody in the presence of 5 U/ml of IL-2.

(CD3). Whereas CD8+ cells proliferated in response to Sepharose (CD3), there was no additional enhancement produced by Sepharose (CD3–CD4).

The effect of soluble anti-CD4 on proliferation induced by immobilized antibodies is shown in Fig. 3. Sepharose (CD3)-

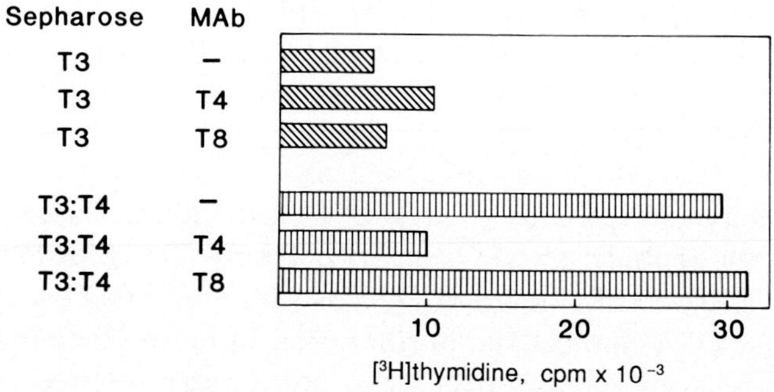

Fig. 3. Effect of soluble antibody on proliferation induced by immobilized antibody: Designated populations of T cells were stimulated by immobilized antibody in the presence of 5 U/ml of IL-2. At the initiation of culture, soluble antisera were added as a 1/5000 dilution of ascites. In all cases, Sepharose-immobilized antibodies were preincubated in media containing 10% human serum to block unoccupied protein A sites, prior to the addition of soluble antisera.

induced proliferation in unseparated T cells was essentially unaffected by soluble anti-CD4 or anti-CD8. The enhanced proliferation induced by Sepharose (CD3–CD4), however, was abrogated in the presence of soluble anti-CD4 but not in the presence of soluble anti-CD8. These results suggest that interference with co-cross-linking between CD3 and CD4 eliminates the proliferative enhancement.

III. Lymphocyte Subpopulations

Monoclonal antibodies reactive with subpopulations of human (2H4+4B4− and 2H4−4B4+) and rat (OX22− and OX22+) CD4+ lymphocytes have been shown to define distinct functional subsets (*7–10*). The CD4+2H4+4B4− population selectively activates CD8 cells to effect suppression of immunoglobulin synthesis in a pokeweed mitogen-stimulated system, resulting in its designation as a suppressor–inducer population (*8*). The reciprocal population, which is phenotypically CD4+2H4−4B4+, functions as a helper cell in immunoglobulin production in the pokeweed mitogen system (*7*). These populations possess distinctive functional programs and differ in their proliferative responses to various stimuli (Table I) (*11*).

The 2H4 antibody recognizes a restricted epitope on the leukocyte common antigen (LCA) family of glycoproteins (*12*). The human LCA gene has been cloned and its genomic organization characterized (*13*). The 5′ region of the LCA gene contains three exons that can be joined in various combinations by alternative splicing. As a result, individual isoforms of LCAs differing in their extracellular domains are produced. Two of the largest LCA isoforms (205 kDa and 220 kDa) include exon 4, which has been shown to encode the 2H4 epitope (*14*). CD4+2H4− lymphocytes expressing these high-molecular-mass LCA isoforms are a functionally distinct lymphocyte subpopulation, and antibodies to these isoforms were shown to block suppression. The reciprocal CD4+2H4− subset only expresses the 180-kDa and 190-kDa isoforms of the LCA family of glycoproteins (*12*).

TABLE I

Response of CD4+ Lymphocyte Subsets to Various Stimuli

Stimulus	Subset response	
	CD4+2H4+	CD4+2H4−
Sepharose (CD3)	+	+
Mitogen	+++	+++
Allo-MLR	+++	+++
Auto-MLR	+++	+
Soluble antigen	±	+++
Sepharose (CD3–CD4)	+	+++
Anti-$T11_2$ + Anti-$T11_3$	+++	+
Help (B/CTL)	−	+++
Suppression induction	+++	+

The differential responsiveness of these lymphocyte subpopulations to soluble antigen presented in the context of self-MHC was of particular interest. It has been postulated that antigen presentation by accessory cells involves the formation of a quaternary complex involving antigen, class II-MHC, CD4, and CD3-Ti (*15*, *16*). If CD3–CD4 cross-linking by Sepharose (CD3–CD4) enhances proliferation by mimicking such a complex, this enhancement might be confined to the T-cell population which normally responds to soluble antigen.

To test this hypothesis, CD4+ lymphocytes were separated into 2H4+ and 2H4− populations by panning. Each of these populations was tested for its proliferative response to immobilized antibodies as shown in Fig. 4. Results from three independent experiments are shown to point out the donor variability in these experiments. Whereas both populations proliferated in response to Sepharose (CD3), enhanced proliferation induced by Sepharose (CD3–CD4) was demonstrated preferentially in the CD4+2H4− helper population. The average stimulation index of Sepharose (CD3 : CD4) was 1.5 ± 0.9 (n = 5) in the CD4+2H4+ population and 4.5 ± 2.4 (n = 5) in the CD4+2H4− population. These

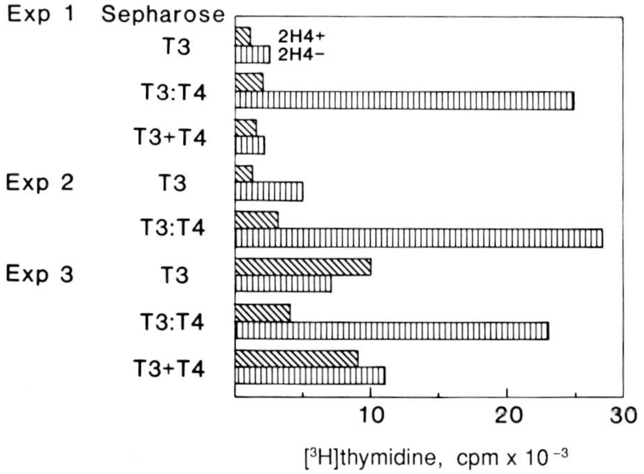

Fig. 4. Proliferative response of T cell subpopulations to immobilized antibodies: CD4+ lymphocytes were separated into 2H4+ and 2H4− populations by adherence to anti-2H4-coated dishes. The proliferative response to Sepharose-immobilized antibody was measured in the presence of 5 U/ml of IL-2.

results suggest that the proliferative enhancement induced by CD3–CD4 cross-linking reflects a preferential activation of the CD4+2H4− helper population. These results suggest that there is a heterobivalent interaction between the antigen–MHC complex on accessory cells and both CD4 and CD3–TCR on the responder cells. This interaction is postulated to induce the aggregation of CD4 with CD3–TCR, an effect mimicked by Sepharose (CD3–CD4). On CD4+2H4− lymphocytes, this coaggregation of CD4 and CD3–TCR results in an activation signal. In contrast, on CD4+2H4+ lymphocytes, either the coaggregation of CD4 and CD3–TCR is precluded, or the signal transduced by this association is not delivered.

IV. Conclusions

The T-cell receptor complex is known to consist of a disulfide-linked heterodimer (Ti) associated noncovalently with three to four transmembrane proteins collectively known as CD3. The aggregation of this complex by immobilized antibodies reactive with either Ti or CD3 is sufficient to induce the proliferation of resting T lymphocytes (*17*). Physiologically, antigen recognition requires specific interactions between T cells and accessory cells bearing self-major histocompatibility molecules. T lymphocytes expressing CD4 recognize antigen in the context of class II MHC whereas CD8-positive lymphocytes are restricted to class I MHC-bearing accessory cells (*18*, *19*). These relationships have been postulated to result from a specific recognition of class II MHC by CD4 and class I MHC by CD8. Such interactions have been further postulated to stabilize T cell–accessory cell adhesion, allowing antigen recognition to occur (*20*). Although such interactions probably play some role in antigen-specific T-cell activation, it is becoming increasingly clear that CD4 and CD8 function as more than just cell adhesion molecules. Perturbation of CD4 by monoclonal antibodies is capable of inhibiting lectin or antibody-induced lymphocyte proliferation (*21*). Furthermore, our results suggest that coaggregation of CD4 and the TCR complex enhances T-lymphocyte activation. The potential for cell surface molecules to interact with one another in the modulation of cellular activation and proliferation may provide an added level of control over lymphocyte growth (*22*).

References

1. J. M. Bishop, *Annu. Rev. Biochem.* **52,** 350–354 (1983).
2. J. Downward, Y. Yarden, E. Mayes, G. Scrace, N. Totty, P. Stockwell, A. Ullrich, and M. D. Waterfield, *Nature* (*London*) **307,** 521–527 (1984).
3. A. F. Williams, *Immunol. Today* **8,** 298–303 (1987).
4. P. Anderson, C. Morimoto, J. B. Breitmeyer, and S. F. Schlossman, *Immunol. Today* **8,** 199–203 (1988).
5. P. Anderson, M. L. Blue, C. Morimoto, and S. F. Schlossman, *J. Immunol.* **139,** 678–682 (1987).
6. P. Anderson, M. L. Blue, C. Morimoto, and S. F. Schlossman, *Cell. Immunol.* **115,** 246–256 (1988).
7. C. Morimoto, N. L. Letvin, A. W. Boyd, *et al., J. Immunol.* **134,** 3762–3769 (1985).

8. C. Morimoto, N. L. Letvin, J. A. Distaso, W. R. Aldrich, and S. F. Schlossman, *J. Immunol.* **134,** 1508–1515 (1985).
9. R. P. Arthur, and D. Mason, *J. Exp. Med.* **163,** 774–786 (1986).
10. A. W. Barclay, D. I. Jackson, A. C. Willis, and A. F. Williams, *EMBO J.* **6,** 1259–1264 (1987).
11. T. Matsuyama, P. Anderson, J. F. Daley, S. F. Schlossman, and C. Morimoto, *Eur. J. Immunol.* **18,** 1473–1476 (1988).
12. C. E. Rudd, C. Morimoto, L. L. Wong, and S. F. Schlossman, *J. Exp. Med.* **166,** 1758–1773 (1987).
13. M. Streuli, L. R. Hall, Y. Saga, S. F. Schlossman, and H. Saito, *J. Exp. Med.* **166,** 1548–1566 (1987).
14. M. Streuli, T. Matsuyama, C. Morimoto, S. F. Schlossman, and H. Saito, *J. Exp. Med.* **166,** 1567–1572 (1987).
15. E. L. Reinherz, S. C. Meuer, and S. F. Schlossman, *Immunol. Rev.* **74,** 83–112 (1983).
16. R. H. Schwartz, *Annu. Rev. Immunol.* **3,** 237–251 (1985).
17. S. C. Meuer, J. C. Hodgdon, R. E. Hussey, J. P. Protentis, S. F. Schlossman, and E. L. Reinherz, *J. Exp. Med.* **158,** 988–996 (1983).
18. S. L. Swain, *Proc. Natl. Acad. Sci. U.S.A.* **78,** 7101–7105 (1981).
19. S. C. Meuer, S. F. Schlossman, and E. L. Reinherz, *Proc. Natl. Acad. Sci. U.S.A.* **79,** 4395–4399 (1982).
20. J. L. Greenstein, B. Malissen, and S. J. Burakoff, *J. Exp. Med.* **162,** 369–374 (1985).
21. I. Bank, and L. Chess. *J. Exp. Med.* **162,** 1294–1301 (1985).
22. P. Anderson, M. L. Blue, and S. F. Schlossman, *J. Immunol.* **140,** 1732–1737 (1988).

6

Transforming Growth Factors

HAROLD L. MOSES, CHARLES C. BASCOM, RUSSETTE M. LYONS, NANCY J. SIPES, AND ROBERT J. COFFEY, JR.

Departments of Cell Biology and Medicine
Vanderbilt University School of Medicine
Nashville, Tennessee

I. Introduction

Transforming growth factors (TGFs) were originally defined by their biological effects on murine fibroblastic cell lines. These effects included induction of morphologic transformation in monolayer culture and stimulation of colony formation in soft agar (for review, see Goustin *et al.*, 1986). The early studies with TGFs indicated that they may play a role in neoplastic transformation and led to the purification and cloning of two very important growth-regulatory molecules, TGFα and TGFβ. Studies with the purified molecules indicate that they probably also play a very important role in normal growth and development. Interestingly, one of these factors, TGFα, is a potent mitogen for most cell

THE REGULATION OF PROLIFERATION AND DIFFERENTIATION IN NORMAL AND NEOPLASTIC CELLS

types while the other, TGFβ, is the most potent growth-inhibitory polypeptide known for most cell types (Tucker *et al.*, 1984a; Moses *et al.*, 1985).

II. TGFα and Its Receptor

First described by DeLarco and Todaro (1978), TGFα was originally termed sarcoma growth factor. It was later shown that the sarcoma growth factor preparation contained both TGFα and TGFβ (Anzano *et al.*, 1983) and that much of the biological activity originally ascribed to sarcoma growth factor was due to the TGFβ. TGFα was purified to homogeneity by Marquardt *et al.* (1984) and is a 50-amino-acid molecule that has some sequence and significant structural identity with epidermal growth factor (EGF). The gene for TGFα has been cloned, indicating a precursor of 160 amino acids that is processed in a complex manner to yield the mature molecule (Derynck *et al.*, 1984; Bringman *et al.*, 1987). TGFα binds to the EGF receptor and appears to mediate all of its biological effects through this EGF receptor binding. No evidence has been presented for a TGFα receptor distinct from the EGF receptor. The cell culture effects of TGFα are virtually identical to those of EGF (Anzano *et al.*, 1983). However, in *in vivo* and organ culture assays, differences between the biologic effects of TGFα and EGF have been observed (for review, see Derynck, 1986). These differences were quantitative, with TGFα tending to be more potent than EGF, and no qualitative differences in the biological effects of EGF and TGFα have been reported.

TGFα was originally shown to be produced by murine sarcoma virus-transformed mouse 3T3 cells but not by the nontransformed parent cells (DeLarco and Todaro, 1978). It was later found in medium conditioned by human carcinoma cells in culture (Todaro *et al.*, 1980). Shortly thereafter, TGFα was identified in embryonic tissue (Twardzik *et al.*, 1982). These initial findings in addition to the absence of reports of finding TGFα in normal adult tissue led to the widespread view that TGFα is an embryonic molecule inappropriately expressed in some cancer cells. It was in conjunction with the discovery of TGFα that the autocrine hypothesis was first published as an explanation for the excessive growth that occurs in neoplastic cells (Sporn and Todaro, 1980).

We have demonstrated that TGFα does occur in normal adult cells which compose the largest organ, the skin. In these studies, secondary cultures of normal keratinocytes were grown in low-calcium, serum-free medium (Wille *et al.*, 1984). These cells require EGF/TGFα for proliferation and retain the ability to differentiate under high-calcium conditions. It was demonstrated that TGFα mRNA and protein are produced by both normal neonatal and adult keratinocytes in culture (Coffey *et al.*, 1987). Additionally, TGFα expression in these cells was found to be dependent upon the presence of EGF; both EGF and TGFα induced significant levels of TGFα mRNA. EGF was demonstrated to stimulate the release of TGFα protein into the culture medium. It was not possible to examine for TGFα induction of protein release because of the exogenously added factor, which would interfere in the ELISA assay. However, TGFα did induce TGFα mRNA, demonstrating autoinduction, and we have speculated that such autoinduction may be a mechanism of signal amplification for finer control of the proliferative response.

Evidence for TGFα mRNA and protein expression *in vivo* was obtained using *in situ* hybridization and immunocytochemistry, respectively (Coffey *et al.*, 1987). This demonstrates that TGFα production does occur in normal adult epithelial cells that are also capable of responding to the factor, suggesting the possibility of autocrine regulation of normal cell proliferation. These studies, along with those from other laboratories demonstrating normal autocrine stimulation by platelet-derived growth factor, insulin-like growth factor-1, and interleukin-2, suggest that the autocrine hypothesis should now be altered (for review, see Goustin *et al.*, 1986). It can no longer be considered a phenomenon occurring only in neoplastic cells resulting in excessive proliferation; rather, it is an important phenomenon involved in the normal regulation of growth control. These findings, however, do not exclude the possibility that abnormal autocrine stimulation by TGFα could be involved in neoplastic transformation. Unregulated production of TGFα could be important in this process. In the case of the normal skin keratinocytes, the production of TGFα is clearly regulated, since the cells produce very little TGFα when grown in the absence of EGF/TGFα for 48 hr (Coffey *et al.*, 1987). However, adequate proof of TGFα autocrine stimulation, such as specific antibody inhibition, has not been provided in cancer cell lines or primary cultures of human neoplasms. Instead of excessive autocrine stimulation, the changes observed in this pathway in neoplastically transformed cells include amplification or deletional mutation of the receptor or changes in the postreceptor signal transduction pathway (for review, see Goustin *et al.*, 1986).

III. TGFβ and Its Receptor

TGFβ was first described as a growth-stimulatory molecule because of its ability to induce soft agar colony formation of mouse embryo-derived fibroblastic AKR2B cells (Moses *et al.*, 1981) and shortly thereafter because of its biological effects in combination with EGF on rat fibroblastic NRK cells (Roberts *et al.*, 1981). Although originally described as being produced by neoplastically transformed cells (Moses *et al.*, 1981; Roberts *et al.*, 1981), TGFβ is now known to be a ubiquitous molecule, and it has been purified from several normal tissues (for review, see Roberts *et al.*, 1983). Platelets, which give rise to the TGFβ found in serum (Childs *et al.*, 1982), are the most abundant source for purification of TGFβ. The intact, active TGFβ molecule has a molecular mass of 25 kDa and is composed of two identical disulfide-linked subunits of 12 kDa each (Assoian *et al.*, 1983). The human gene for TGFβ has been cloned and the amino acid sequence deduced from the cDNA sequence (Derynck *et al.*, 1985). This indicates a subunit of 112 amino acids and suggests a precursor encoded in a 390-residue open reading frame. The precursor is processed by proteolytic cleavage to yield the active molecule (Gentry *et al.*, 1987). Murine

TGFβ has also been cloned, and comparison with the human sequence shows an exceptionally high degree of evolutionary conservation (Derynck *et al.*, 1986).

A second TGFβ (called TGFβ2 to distinguish it from the originally described TGFβ, now called TGFβ1) has been identified in porcine platelets and bovine bone (Cheifetz *et al.*, 1987). TGFβ2 apparently binds to the same receptor as TGFβ1 and has virtually identical biological activities (Cheifetz *et al.*, 1987; H. L. Moses, C. C. Bascom, R. M. Lyons, and N. J. Sipes, unpublished observations). An apparently identical molecule has been identified as an immunosuppressive agent produced by glioma cells (Wrann *et al.*, 1987). The growth inhibitor from African Green monkey kidney cells, BSC-1, originally described by Holley *et al.* (1978) has now been cloned and the cDNA sequence indicates that it is identical to TGFβ2 (Hanks *et al.*, 1988). The term *polyergin* has been suggested by these authors to distinguish this molecule from the TGFβ isolated from human platelets. Other molecules with structural and some sequence similarity to TGFβ have been purified or identified by gene cloning and DNA sequencing. These include Müllerian inhibiting substance (Cate *et al.*, 1986), inhibins (and their β-chain dimers, activins) (Mason *et al.*, 1985), and the *Drosophila* decapentaplegic gene (Padgett *et al.*, 1987). Müllerian inhibiting substance and inhibins and activins have receptors that are distinct from the TGFβ receptor (Ying *et al.*, 1986; Coughlin *et al.*, 1987).

TGFβ has specific cell surface receptors which, like the TGFβ molecule itself, are ubiquitous (Tucker *et al.*, 1984b). Specific binding of ^{125}I-labeled TGFβ to various mesenchymal and epithelial cells in primary and secondary culture and continuous cell lines, both normal and neoplastic, has been reported. The dissociation constants reported have ranged from 25 to 140 p*M* and receptor number per cell from 10,000 to 40,000 (Frolik *et al.*, 1984; Tucker *et al.*, 1984b; Massague and Like, 1985). The TGFβ receptor is apparently quite different from other growth factor receptors. At least two types of receptors for TGFβ have been proposed on the basis of chemical cross-linking studies (Massague, 1985; Cheifetz *et al.*, 1987). No kinase or other enzymatic activities have been reported for the TGFβ receptor thus far.

TGFβ has been reported to have numerous and diverse biological activities. It is mitogenic only for fibroblastic and selected other mesenchymal cells (Moses *et al.*, 1985). Data have been presented indicating that TGFβ mitogenesis in monolayer culture is indirect through induction of c-*sis* and autocrine stimulation by endogenous platelet-derived growth factor (Leof *et al.*, 1986). TGFβ stimulation of soft agar growth has been shown to be secondary to induction of fibronectin synthesis and release (Ignotz and Massague, 1986). TGFβ stimulates extracellular matrix production by increasing synthesis of matrix components including procollagen type I and fibronectin (Ignotz and Massague, 1986; Ignotz *et al.*, 1987; Raghow *et al.*, 1987), by diminishing matrix degradation through stimulating protease inhibitor (Laiho *et al.*, 1986), and by decreasing protease production (Matrisian *et al.*, 1986; Edwards *et al.*, 1987). TGFβ also is a very potent chemotactic factor for dermal fibroblasts (Postlethwaite *et al.*, 1987). All of these actions probably contribute to the ability of TGFβ to stimulate connective tissue formation *in vivo* (Roberts *et al.*, 1986). TGFβ also inhibits differentiation of adipocytes and myoblasts (Massague *et al.*, 1987).

We demonstrated that the growth inhibitor originally described by Holley *et al.* (1978) from BSC-1 cells is similar to human platelet-derived TGFβ (Tucker *et al.*, 1984a). Holley *et al.* (1978) have suggested that autocrine inhibition by polyergin was important in the growth regulation of BSC-1 cells. In addition to demonstrating biochemical similarities, it was further shown that human platelet-derived TGFβ, like polyergin, is a highly potent inhibitor of BSC-1 and CCL-64 (mink lung) epithelial cells. This led to studies of the inhibitory effects of TGFβ on a variety of normal cells demonstrating that TGFβ is the most potent growth inhibitory polypeptide known for a wide variety of cell types including epithelial, lymphoid, and myeloid cells (for review, see Moses and Leof, 1986).

We have also demonstrated that TGFβ is a potent growth inhibitor for secondary cultures of human foreskin keratinocytes (Moses *et al.*, 1985; Shipley *et al.*, 1986). The keratinocytes were found to be reversibly inhibited in their growth by TGFβ with the majority of cells blocked in the G_1 phase of the cell cycle. Half-maximal inhibition was obtained at 12 p*M* TGFβ. There was no induction of any of several differentiation markers examined, indicating that the mechanism of growth inhibition is not through induction of terminal differentiation. The keratinocytes were also demonstrated to synthesize and release TGFβ into the medium, with confluent cultures producing as much as 80 p*M* per 24 hr (Shipley *et al.*, 1986). However, all the detectable TGFβ released was in a latent form detectable only after acid treatment of the medium, similar to observations of many other cell types (Lawrence *et al.*, 1984). Whether the latent TGFβ activates spontaneously or can be activated by the cells, with subsequent binding to cell surface receptors, is not known. However, since the keratinocytes have receptors for TGFβ (Shipley *et al.*, 1986), are capable of responding to the factor, and secrete relatively large quantities into conditioned medium, the possibility of negative autocrine regulation by TGFβ in normal keratinocytes seems highly probable. TGFβ is released by cells in culture (including the human keratinocytes) and by platelets in a latent form that is irreversibly activated by acid treatment (Lawrence *et al.*, 1984). Various treatments which dissociate hydrogen bonds, such as high salt concentrations, urea, and detergents, also result in irreversible activation of TGFβ (Pircher *et al.*, 1986; R. M. Lyons and H. L. Moses, unpublished observations). These and other data suggest that the activation of the latent form is separate from the proteolytic processing steps known to be necessary to generate the 25-kD molecule (Gentry *et al.*, 1987). Processing appears to have occurred already by the time TGFβ is released from the cells. Recent studies on possible physiological mechanisms of activation have demonstrated that plasmin, a ubiquitous serine protease, can activate at least a portion of the TGFβ released by cells in culture (Lyons *et al.*, 1988). Plasmin activation of TGFβ as a physiological mechanism is attractive because of the wide distribution of plasminogen *in vivo* and because it has been shown that TGFβ induces the endothelial-cell-type plasminogen-activator inhibitor, PAI-1 (Laiho *et al.*, 1986). This could provide a negative feedback control of TGFβ activation.

IV. Autocrine Regulation of Epithelial Cells by TGFs

The data summarized above demonstrate that control of keratinocyte proliferation may involve both positive and negative polypeptide regulators that bind to cell surface membrane receptors. The keratinocytes produce the same peptides that stimulate or inhibit their proliferation. TGFα stimulates proliferation and the keratinocytes produce TGFα: this production is autoregulated (Fig. 1). It is postulated that such autoregulation may provide needed amplification of the growth-stimulatory signal. TGFβ reversibly inhibits proliferation of keratinocytes, cells that also secrete this factor in a latent form (Fig. 1). The latent material could be activated in the medium or at the cell surface by proteolytic action, resulting in a negative effect on cell proliferation. Other growth-inhibitory polypeptides such as the interferons (IFN) could also be involved (Fig. 1). It is hypothesized that autocrine regulation by both stimulators and inhibitors of proliferation is an important

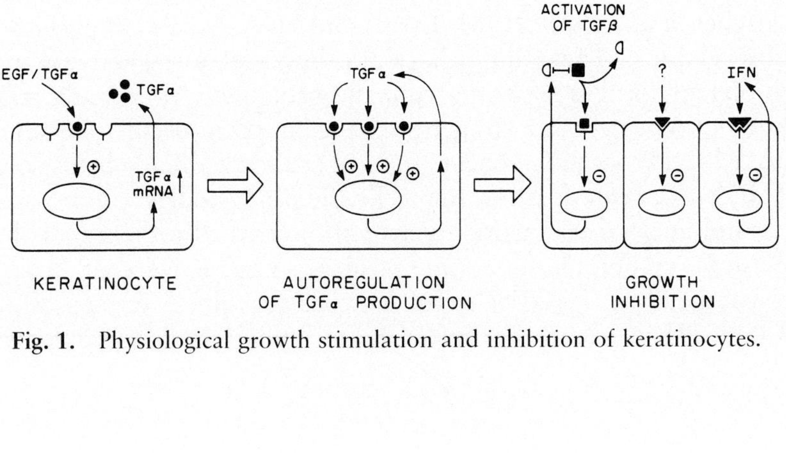

Fig. 1. Physiological growth stimulation and inhibition of keratinocytes.

physiological mechanism of regulation of keratinocyte proliferation. A similar mechanism may be involved in the regulation of proliferation of other cell types as well and may involve peptide growth factors and growth inhibitors other than TGFα and TGFβ. The presence of opposing regulatory pathways would allow for a more precise control of the very important process of cell proliferation than an on–off stimulatory pathway provided by the growth factors. This suggests that the growth-inhibitory polypeptides may play as important a role in the control of cell proliferation as the growth-stimulatory factors.

V. Changes in Autocrine Regulation in Neoplastic Transformation

Alterations in the TGFα autocrine stimulatory pathway in neoplastic transformation have been proposed (Todaro *et al.*, 1980; Coffey *et al.*, 1987) and could result in an increased proliferative potential. However, changes in this pathway that result in increased stimulation seem to involve either increased numbers of receptors or deletional mutations causing constitutive activation in the receptor pathway, rather than increased autocrine stimulation (Fig. 2). The role of TGFβ in neoplastic transformation of epithelial cells is probably very different from that involved in fibroblastic cells, in which autocrine stimulation may occur (Keski-Oja *et al.*, 1987). We have demonstrated that a squamous carcinoma cell line has lost the inhibitory response to TGFβ exhibited by normal keratinocytes (Shipley *et al.*, 1986) and similar results have been obtained by Masui *et al.* (1986) in bronchial-derived squamous carcinoma cell lines. Both loss of receptor (Fig. 2, step 2) and loss of the normal inhibitory response in the presence of apparently normal receptor binding (Fig. 2, step 3) have been reported (Shipley *et al.*, 1986; Masui *et al.*, 1986). Loss of the ability to activate the latent TGFβ released by most cells, including platelets, could also result in loss of the normal inhibitory effect of TGFβ (Fig. 2, step 1). Both increased autocrine

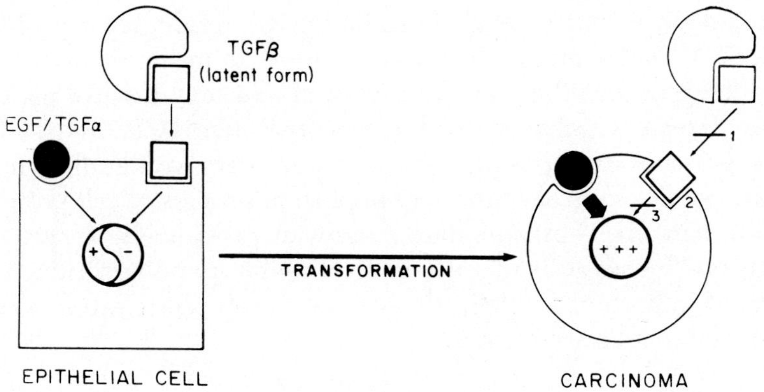

Fig. 2. Changes in TGF autocrine regulation in neoplastic transformation.

stimulation by TGFα or decreased autocrine inhibition by TGFβ could lead to an increased proliferative potential and thereby contribute to the neoplastic phenotype.

References

Anzano, M. A., Roberts, A. B., Smith, J. M., Sporn, M. B., and DeLarco, J. E. (1983). Sarcoma growth factor from conditioned medium of virally transformed cells is composed of both type alpha and type beta transforming growth factors. *Proc. Natl. Acad. Sci. U.S.A.* **80**, 6264–6268.

Assoian, R. K., Komoriya, A., Meyers, C. A., Miller, D. M., and Sporn, M. B. (1983). Transforming growth factor-beta in human platelets. Identification of a major storage site, purification, and characterization. *J. Biol. Chem.* **258**, 7155–7160.

Bringman, T. S., Lindquist, P. M., and Derynck, R. (1987). Different transforming growth factor-alpha species are derived from a glycosolated and palmitoylated transmembrane precursor. *Cell* (*Cambridge, Mass.*) **48**, 429–440.

Cate, R. L., Mattaliano, R. J., Hession, C., Tizard, R., Farber, N. M., Cheung, A., Ninfa, E. G., Frey, A. Z., Gash, D. J., Chow, E. P., Fisher, R. A., Bertonis, J. M., Torres, G., Wallner, B. P., Ramachandran, K. L., Ragin, R. C., Managanaro, T. F., MacLaughlin, D. T., and Donahoe, P. K. (1986). Isolation of the bovine and human genes for Müllerian inhibiting substance and expression of the human gene in animal cells. *Cell (Cambridge, Mass.)* **45**, 685–698.

Cheifetz, S., Weatherbee, J. A., Tsang, M. L.-S., Anderson, J. K., Mole, J. E., Lucas, R., and Massague, J. (1987). The transforming growth factor-beta system, a complex pattern of cross-reactive ligands and receptors. *Cell (Cambridge, Mass.)* **48**, 409–415.

Childs, C. B., Proper, J. A., Tucker, R. F., and Moses, H. L. (1982). Serum contains a platelet-derived transforming growth factor. *Proc. Natl. Acad. Sci. U.S.A.* **79**, 5312–5316.

Coffey, R. J., Derynck, R., Wilcox, J. N., Bringman, T. S., Goustin, A. S., Moses, H. L., and Pittelkow, M. R. (1987). Production and auto-induction of transforming growth factor-alpha in human keratinocytes. *Nature (London)* **328**, 817–820.

Coughlin, J. P., Donahoe, P. K., Budzik, G. P., and MacLaughlin, D. T. (1987). Müllerian inhibiting substance blocks autophosphorylation of the EGF receptor by inhibiting tyrosine kinase. *Mol. Cell. Endocrinol.* **49**, 75–86.

DeLarco, J. E., and Todaro, G. J. (1978). Growth factors from murine sarcoma virus-transformed cells. *Proc. Natl. Acad. Sci. U.S.A.* **75**, 4001–4005.

Derynck, R. (1986). Transforming growth factor-alpha: Structure and biological activities. *J. Cell. Biochem.* **32**, 293–304.

Derynck, R., Roberts, A. B., Winkler, M. E., Chen, E. Y., and Goeddel, D. V. (1984). Human transforming growth factor-alpha: Precursor structure and expression in *E. coli*. *Cell (Cambridge, Mass.)* **38**, 287–297.

Derynck, R., Jarrett, J. A., Chen, E. Y., Eaton, D. H., Bell, J. R., Assoian, R. K., Roberts, A. B., Sporn, M. B., and Goeddel, D. V. (1985). Human transforming growth factor-beta complementary DNA sequence and expression in normal and transformed cells. *Nature (London)* **316**, 701–705.

Derynck, R., Jarrett, J. A., Chen, E. Y., and Goeddel, D. V. (1986). The murine transforming growth factor-beta precursor. *J. Biol. Chem.* **261**, 4377–4379.

Edwards, D. R., Murphy, G., Reynolds, J. J., Whithman, S. E., Docherty, A. J. P., Angel, P., and Heath, J. K. (1987). Transforming growth factor beta modulates the expression of collagenase and metalloproteinase inhibitor. *EMBO J.* **6**, 1899–1904.

Frolik, C. A., Wakefield, L. M., Smith, D. M., and Sporn, M. B. (1984). Characterization of a membrane receptor for transforming growth factor-beta in normal rat kidney fibroblasts. *J. Biol. Chem.* **259**, 10995–11000.

Gentry, L. E., Webb, N. R., Lim, G. J., Brunner, A. M., Ranchalis, J. E., Twardzik, D. R., Lioubin, M. N., Marquardt, H., and Purchio, A. F. (1987). Type I transforming growth factor beta: Amplified expression and secretion of mature and precursor polypeptides in Chinese hamster ovary cells. *Mol. Cell. Biol.* **7**, 3418–3427.

Goustin, A. S., Leof, E. B., Shipley, G. D., and Moses, H. L. (1986). Perspectives in cancer research: Growth factors and cancer. *Cancer Res.* **46,** 1015–1029.

Hanks, S. K., Armour, R., Baldwin, J. H., Maldonado, F., Spiess, J., and Holley, R. W. (1988). Amino acid sequence of the BSC-1 cell growth inhibitor (polyergin) deduced from the nucleotide sequence of the cDNA. *Proc. Natl. Acad. Sci. U.S.A.* **85,** 79–82.

Holley, R. W., Armour, R., and Baldwin, J. H. (1978). Density-dependent regulation of growth of BSC-1 cells in cell culture: Growth inhibitors formed by the cells. *Proc. Natl. Acad. Sci. U.S.A.* **75,** 1864–1866.

Ignotz, R. A., and Massague, J. (1986). Transforming growth factor-beta stimulates the expression of fibronectin and collagen and their incorporation into the extracellular matrix. *J. Biol. Chem.* **261,** 4337–4345.

Ignotz, R. A., Endo, R., and Massague, J. (1987). Regulation of fibronectin and type I collagen mRNA levels by transforming growth factor-beta. *J. Biol. Chem.* **262,** 6443–6446.

Keski-Oja, J., Lyons, R. M., and Moses, H. L. (1987). Immunodetection and modulation of cellular growth with antibodies against native transforming growth factor-beta. *Cancer Res.* **47,** 6451–6458.

Laiho, M., Saksela, O., Andreasen, P. A., and Keski-Oja, J. (1986). Enhanced production and extracellular deposition of the endothelial-type plasminogen activator inhibitor in cultured human lung fibroblasts by transforming growth factor-beta. *J. Cell Biol.* **103,** 2403–2410.

Lawrence, D. A., Pircher, R., Kryceve-Martinerie, C., and Jullien, P. (1984). Normal embryo fibroblasts release transforming growth factors in a latent form. *J. Cell. Physiol.* **121,** 184–188.

Leof, E. B., Proper, J. A., Goustin, A. S., Shipley, G. D., DiCorleto, P. E., and Moses, H. L. (1986). Induction of c-*sis* mRNA and activity similar to platelet-derived growth factor by transforming growth factor-beta: A proposed model for indirect mitogenesis involving autocrine activity. *Proc. Natl. Acad. Sci. U.S.A.* **83,** 2453–2457.

Lyons, R. M., Keski-Oja, J., and Moses, H. L. (1988). Proteolytic activation of latent transforming growth factor-beta from fibroblast conditioned medium. *J. Cell Biol.* **106,** 1659–1665.

Marquardt, H., Hunkapiller, M. W., Hood, L. E., and Todaro, G. J. (1984). Rat transforming growth factor type 1: Structure and relation to epidermal growth factor. *Science* **223,** 1079–1082.

Mason, A. J., Hayflick, J. S., Ling, N., Esch, F., Ueno, N., Ying, Y., Guillemin, R., Niall, H., and Seeburg, P. H. (1985). Complementary DNA sequences of ovarian follicular fluid inhibin show precursor structure and homology with transforming growth factor-beta. *Nature* (*London*) **318,** 659–663.

Massague, J. (1985). Subunit structure of a high-affinity receptor for type beta-transforming growth factor. Evidence for a disulfide-linked glycosylated receptor complex. *J. Biol. Chem.* **260,** 7059–7066.

Massague, J., and Like, B. (1985). Cellular receptors for type beta transforming growth factor. Ligand binding and affinity labeling in human and rodent cell lines. *J. Biol. Chem.* **260,** 2636–2645.

Masui, T., Wakefield, L. M., Lechner, J. F., LaVeck, M. A., Sporn, M. B., and Harris, C. C. (1986). Type beta transforming growth factor is the primary differentiation-inducing serum factor for normal human bronchial epithelial cells. *Proc. Natl. Acad. Sci. U.S.A.* **83,** 2438–2442.

Matrisian, L. M., Leroy, P., Ruhlmann, C., Gesnel, M.-C., and Breathnach, R. (1986). Isolation of the oncogene and epidermal growth factor-induced transin gene: Complex control in rat fibroblasts. *Mol. Cell. Biol.* **6,** 1679–1686.

Moses, H. L., and Leof, E. B. (1986). Transforming growth factor beta. *In* "Oncogenes and Growth Control" (P. Kahn and T. Graf, eds.), pp. 51–57. Springer-Verlag, Berlin and New York.

Moses, H. L., Branum, E. B., Proper, J. A., and Robinson, R. A. (1981). Transforming growth factor production by chemically transformed cells. *Cancer Res.* **41,** 2842–2848.

Moses, H. L., Tucker, R. F., Leof, E. B., Coffey, R. J., Halper, J., and Shipley, G. D. (1985). Type beta transforming growth factor is a growth stimulator and a growth inhibitor. *Cancer Cells* **3,** 65–71.

Padgett, R. W., St. Johnston, R. D., and Gelbart, W. M. (1987). A transcript from a *Drosophila* pattern gene predicts a protein homologous to the transforming growth factor-beta family. *Nature (London)* **325,** 81–84.

Pircher, R., Jullien, P., and Lawrence, D. A. (1986). Beta-transforming growth factor is stored in human blood platelets as a latent high molecular weight complex. *Biochem. Biophys. Res. Commun.* **136,** 30–37.

Postlethwaite, A. E., Keski-Oja, J., Moses, H. L., and Kang, A. H. (1987). Stimulation of the chemotactic migration of human fibroblasts by transforming growth factor beta. *J. Exp. Med.* **165,** 251–256.

Raghow, R., Postlethwaite, A. E., Keski-Oja, J., Moses, H. L., and Kang, A. H. (1987). Transforming growth factor-beta increases steady-state levels of type I procollagen and fibronectin mRNAs posttranscriptionally in cultured human dermal fibroblasts. *J. Clin. Invest.* **79,** 1285–1288.

Roberts, A. B., Anzano, M. A., Lamb, L. C., Smith, J. M., and Sporn, M. B. (1981). New class of transforming growth factors potentiated by epidermal growth factor: Isolation from non-neoplastic tissues. *Proc. Natl. Acad. Sci. U.S.A.* **78,** 5339–5343.

Roberts, A. B., Frolik, C. A., Anzano, M. A., and Sporn, M. B. (1983). Transforming growth factors from neoplastic and non-neoplastic tissues. *Fed. Proc. Fed. Am. Soc. Exp. Biol.* **42,** 2621–2626.

Roberts, A. B., Sporn, M. B., Assoian, R. K., Smith, J. M., Roche, N. S., Wakefield, L. M., Heine, U. I., Liotta, L. A., Falanga, V., Kehrl, H., and Fauci, A. S. (1986). Transforming growth factor type beta: Rapid induction of fibrosis and angiogenesis in vivo and stimulation of collagen formation in vitro. *Proc. Natl. Acad. Sci. U.S.A.* **83,** 4167–4171.

Shipley, G. D., Pittelkow, M. R., Wille, J. J., Scott, R. E., and Moses, H. L. (1986). Reversible inhibition of normal human prokeratinocyte proliferation by type beta transforming growth factor-growth inhibitor in serum-free medium. *Cancer Res.* **46,** 2068–2071.

Sporn, M. B., and Todaro, G. J. (1980). Autocrine secretion and malignant transformation of cells. *N. Engl. J. Med.* **303,** 878–880.

Todaro, G. J., Fryling, C., and DeLarco, J. E. (1980). Transforming growth factors produced by certain human tumor cells: Polypeptides that interact with epidermal growth factor receptors. *Proc. Natl. Acad. Sci. U.S.A.* **77,** 5258–5262.

Tucker, R. F., Shipley, G. D., Moses, H. L., and Holley, R. W. (1984a). Growth inhibitor from BSC-1 cells closely related to the platelet type beta transforming growth factor. *Science* **226,** 705–707.

Tucker, R. F., Branum, E. L., Shipley, G. D., Ryan, R. J., and Moses, H. L. (1984b). Specific binding to cultured cells of ^{125}I-labeled transforming growth factor-type beta from human platelets. *Proc. Natl. Acad. Sci. U.S.A.* **81,** 6757–6761.

Twardzik, D. R., Ranchalis, J. E., and Todaro, G. J. (1982). Mouse embryonic transforming growth factors related to those isolated from tumor cells. *Cancer Res.* **42,** 590–593.

Wille, J. J., Pittlekow, M. R., Shipley, G. D., and Scott, R. E. (1984). Integrated control of growth and differentiation of normal human prokeratinocytes cultured in serum-free medium: Clonal analysis, growth kinetics, and cell cycle studies. *J. Cell. Physiol.* **121,** 31–44.

Wrann, M., Bodmer, S., de Martin, R., Siepl, C., Hofer-Warbinek, R., Frei, K., Hofer, E., and Fontana, A. (1987). T cell suppressor factor from human glioblastoma cells is a 12.5-kd protein closely related to transforming growth factor-beta. *EMBO J.* **6,** 1633–1636.

Ying, S.-Y., Becker, A., Ling, N., Ueno, N., and Guillemin, R. (1986). Inhibin and beta type transforming growth factor (TGF beta) have opposite modulating effects on the follicle stimulating hormone (FSH)-induced aromatase activity of cultured rat granulosa cells. *Biochem. Biophys. Res. Commun.* **136,** 969–975.

PART III

The Transformed Cell

7

Transformation of Human Cells by Epstein–Barr Virus

SAMUEL H. SPECK AND JACK L. STROMINGER

Dana-Farber Cancer Institute
Harvard Medical School
Boston, Massachusetts

I. Introduction

EBV (Epstein–Barr virus) was the first human virus discovered that was associated with human tumor development. It is associated with the development of both African Burkitt's lymphoma and nasopharyngeal carcinoma. In addition, it is clearly the etiologic agent of a widespread disease called infectious mononucleosis. In fact, virtually every adult individual has been infected with Epstein–Barr virus and produces it periodically in his or her throat. Most individuals are infected prior to puberty in which case the symptoms are very muted (e.g., a weekend fever).

Infection after puberty, however, results in infectious mononucleosis, a self-limited lymphoproliferation induced by the virus. The mechanism of regulation of this lymphoproliferation is itself a subject of considerable interest.

The discovery of the lymphoma was largely due to the efforts of Dennis Burkitt, a British physician working in Africa, who noticed in his travels through Africa that jaw tumors in children had an unusually high incidence. The exact reason that the tumor appears predominantly as a jaw tumor is not clear, but one possibility is that the parotid duct epithelium is a primary tissue in which this virus replicates. The tumor can also, however, appear as a primary lesion in other locations, for example, in the abdomen. The tumor in some cases is remarkably sensitive to chemotherapy (for background, see Ref. *1* and *2*).

Burkitt also mapped the geographical distribution of the disease. He found that its geographical distribution in Africa corresponded to the malaria zone, and postulated therefore that the disease might be caused by an infectious agent. An interesting question is whether in Africa the disease appears primarily as a response to an initial infection or only develops after a long period of infection. A prospective study of the development of this disease in children in Uganda was carried out by de Thé and colleagues (*3*) under the most unimaginably difficult conditions. Serum samples from a very large number of children were collected and later used to assess the development of tumors in these children along with the development of seropositivity from the collected serum samples. Seropositivity under those relatively primitive socioeconomic conditions occurred in virtually 100% of children by age two, but tumor incidence reached a peak later, at age 6 or 7. It is, therefore, clear that the disease is not a response to an initial primary infection with the virus but develops only after a long period of infection.

A second cancer associated with this virus is nasopharyngeal carcinoma. This association was initially discovered through serological studies, i.e., patients with the disease were found to have high titers of anti-EBV antibodies (*4*). This disease occurs primarily in middle age and has a particularly high incidence in Chinese of Cantonese origin. It is a major disease along the southeast coast of China. It often appears initially as metastatic lesions in the cervical lymph nodes.

The next important step occurred when Epstein obtained samples of tissues from patients with Burkitt's lymphoma, examined them under the microscope and saw nothing. However, when he cultured some of these tissues, he discovered the virus which now bears his name and that of his technician, Yvonne Barr. For their pioneering work, Burkitt and Epstein were early recipients of the Bristol-Myers Award in cancer research.

What is the mechanism of growth transformation by this virus? The virus will produce growth transformation of B lymphocytes infected *in vitro*. However, these cells, which grow permanently in culture, are nevertheless not oncogenic in the sense that they do not cause tumors in nude mice. Tumor development involves a chromosomal translocation involving and activating the *myc* gene on chromosome 8 and one of the three *Ig* loci on chromosomes 2 (λ locus), 14 (*H* locus), or 22 (κ locus). What do the growth-transforming genes of the Epstein–Barr virus contribute to the final tumor development? It seems likely that the virus, which has such profound growth-transforming potential for B lymphocytes, makes some important contribution to the final oncogenic state. In fact, introduction of a deregulated *myc* gene into transgenic animals induces a clonal leukemia, *not* a generalized leukemogenesis (5, 6). Thus, it is clear that the deregulated *myc* gene present in all of the lymphocytes in the transgenic mice does not induce leukemogenesis. The clones of leukemia cells which are produced must then have undergone a second event leading to leukemia, presumably a cellular genetic mutation, which might also be provided by an EBV-encoded gene in Burkitt's lymphoma.

A second interesting property of this virus is that it produces a latent infection of B lymphocytes with little or no virus production. How is latency maintained by this virus? The paradigm of a latent virus is the *Escherichia coli* phage λ. To what extent is that paradigm reflected in the mechanism by which latency is maintained in Epstein-Barr virus infections? The mechanism of latency is only beginning to be studied (*7*).

II. EBV DNA, Transcripts, Proteins, and Their Functions

A. DNA

The first problem was to obtain viral DNA, not an easy job with a virus which is primarily latent; very few viral particles can be obtained, even from "producer" lines. This problem was overcome by cloning the *Bam*HI fragments of the virus and establishing a linkage map of these fragments (*8, 9*) (Fig. 1). The complete DNA sequence of the virus was obtained (*10*) and has been invaluable to many investigators in subsequent studies.

Another problem with this virus is that the analysis of the functions of the genes of most viruses is carried out by mutational means. However, with a virus which is not lytic, one cannot carry out mutational analysis in the usual way. So far, only one nontransforming mutant of EBV has been identified. It was called the P3HR-1 strain of EBV and subsequently found by several laboratories to carry a deletion in the *Bam*HI Y and H fragments. Later, the nontransforming phenotype was mapped to the deletion by recombinational analysis (*11*). Thus, at least one gene important in transformation is encoded in the region of the wild-type virus which is deleted in that mutant.

B. Transcription

The initial latent transcript from the Epstein–Barr virus covers about 100 kb of the 180 kb of the viral genome (*12*) (Fig. 2). It starts at a promoter in *Bam*HI W at the first repeat and terminates in *Bam*HI K. A second promoter slightly 5′ of the first in *Bam*HI C can also be used. Its usage may distinguish to some extent cell lines transformed with EBV *in vitro* and Burkitt's lymphoma cells. The initial transcript can be spliced in many ways to make transcripts

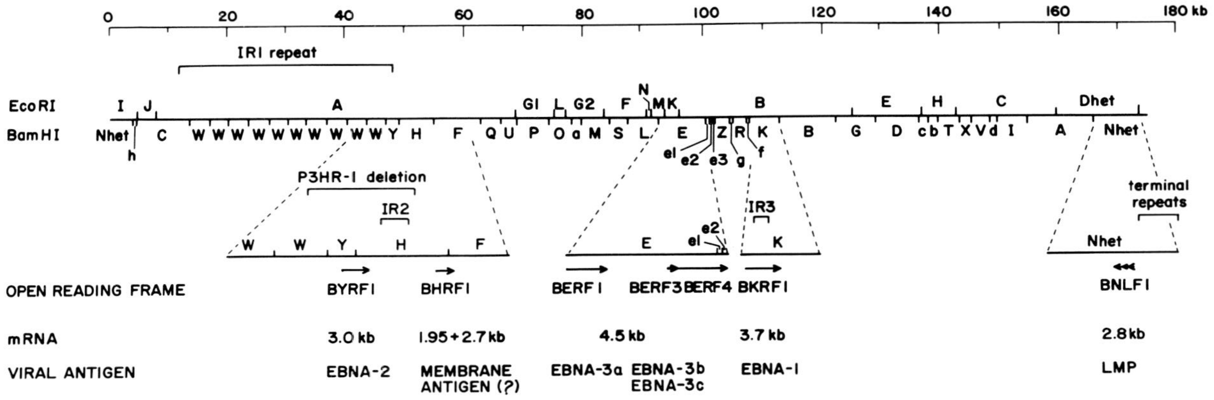

Fig. 1. *Bam*HI and *Eco*RI restriction endonuclease maps of the B95.8 strain cf EBV. Regions of the viral genome that are transcribed in latently infected lymphocytes are shown. The relevant open reading frames and the viral proteins they encode are indicated. Abbreviations: EBV, Epstein-Barr virus; EBNA, Epstein-Barr nuclear antigen; BERF, *Bam*HI E rightward open reading frame; BHRF, *Bam*HI H rightward open reading frame; BKRF, *Bam*HI K rightward open reading frame; BYRF, *Bam*HI Y rightward open reading frame; BMLF, *Bam*HI M leftward open reading frame; BNLF, *Bam*HI N leftward open reading frame; BZLF, *Bam*HI Z leftward open reading frame. EBNA-4 (not shown) is a protein encoded in the *Bam* HI W repeats (see Fig. 2).

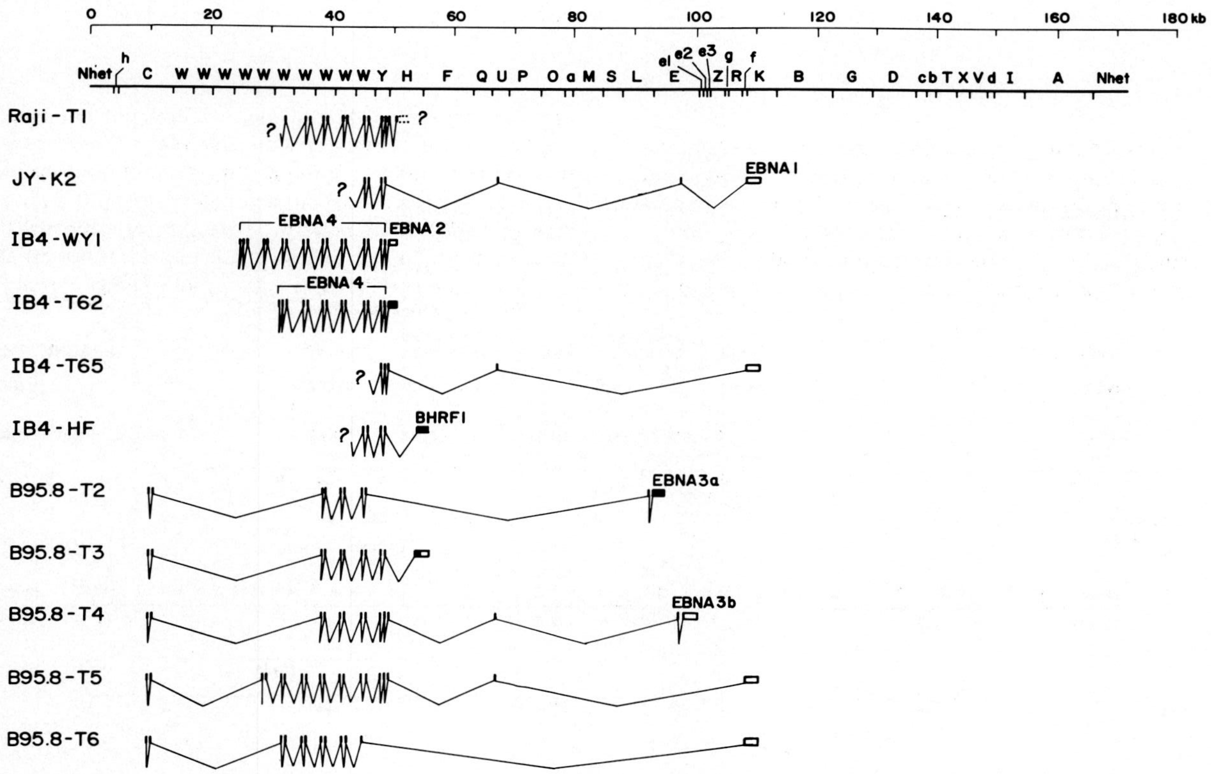
0
20
40
60
80
100
120
140
160
180 kb
Nhet h C W W W W W W W W W W Y H F Q U P O a M S L E e1 e2 e3 g f Z R K B G D c b T X V d I A Nhet
Raji - T1
JY - K2
EBNA 1
IB4 - WY1
EBNA 4
EBNA 2
IB4 - T62
EBNA 4
IB4 - T65
IB4 - HF
BHRF1
B95.8 - T2
EBNA3a
B95.8 - T3
B95.8 - T4
EBNA3b
B95.8 - T5
B95.8 - T6

from which the six known EBV-specific nuclear proteins are translated. Several additional transformation-specific transcripts have been described (*13, 14*).

C. Proteins

Six *nuclear* proteins are made from those spliced mRNAs, EBNA-1, -2, -3a, -3b, -3c, and -4 (Figs. 1 and 2). At least one and possibly three *membrane* proteins are also made in the transformed state. One of the latter is encoded at the *Bam*HI H–F junction and is made from the long rightward transcript (*13*). The only latent protein which is transcribed in the leftward direction in the transformed state is the protein called LMP (latent membrane protein), which is encoded in the terminal Nhet fragment (*15*). All the other latent proteins are transcribed from a latent promoter located in the *Bam*HI W and/or C fragments. The six nuclear proteins are first of all, EBNA-1, described many years ago by Reedman and Klein (16) through immunofluorescence studies. It is encoded in the *Bam*HI K fragment. EBNA-2 is a very important protein, because it is encoded in the *Bam*HI Y and H fragments which are deleted in the nontransforming mutant. EBNAs -3a, -3b, and -3c are encoded in the *Bam*HI E fragment, and EBNA-4 is encoded primarily by the *Bam*HI W repeats (*12*). An additional transcript arising from the circularized viral genome, within Nhet, may encode another membrane protein (14).

Fig. 2. Schematic exon maps of rightward viral transcripts are shown with respect to the *Bam*HI restriction endonuclease map of the B95.8 strain of EBV. Filled boxes represent sequences present in each cDNA clone, and open boxes represent proposed structures. The coding sequences for the known viral antigens present in latently infected lymphocytes are indicated. The cDNA clones are named according to the cell line from which they were derived (Raji, a latently infected Burkitt's lymphoma cell line; JY and IB4, latently infected lymphoblastoid cell lines; B95.8, productively infected marmoset lymphoblastoid cell line) (*12*).

D. Function

An increasing amount of information is available about the transcripts and proteins synthesized in the latent state, but virtually nothing is known about the functions of these proteins in inducing growth transformation. EBNA-1 is in many ways analogous to the SV-40 T antigen. It is required for replication of the virus and it binds to OriP, the origin of replication, which is located in the *Bam*HI C fragment (*17*). It is interesting that OriP and the latent promoters in *Bam*HI C and W have a similar genetic organization to the SV-40 origin of replication and the promotor for transcription of early SV-40 genes. EBNA-2 induces the expression of CD23, the $F_c\epsilon$ receptor of human B cells (*18*), and both EBNA-2 and the latent membrane protein encoded in *Bam*HI Nhet are able to produce morphological growth transformation of RAT-1 cells (19, 20). Nothing else is presently known about the functions of these eight or nine latent proteins.

Thus, the mechanism of transformation of human cells by EBV has been shown to be an enormously complex process. Tools for its analysis are now available, and its study will be rewarding both in terms of our knowledge of molecular virology and of our understanding of oncogenesis in human cells.

References

1. M. A. Epstein and A. G. Achong, eds., "The Epstein-Barr Virus: Recent Advances". Wiley, New York, 1986.
2. D. P. Burkitt and D. H. Wright, eds., "Burkitt's Lymphoma". Livingstone, Edinburgh, 1970.
3. G. de Thé, E. Geser, N. E. Day, P. M. Tukei, E. H. Williams, D. P. Beri, P. G. Smith, A. G. Dean, G. W. Bornkamm, P. Feorino, and W. Henle, *Nature* (*London*) **274,** 756 (1978).
4. L. J. Old, E. A. Boyse, H. F. Wettgen, E. deHarven, G. Geering, B. Williamson, and P. Cufford *Proc. Natl. Acad. Sci. U.S.A.* **56,** 1699 (1966).
5. J.M. Adams, A. W. Harris, C. A. Pinkert, L. W. Corcoran, W. S. Alexander, S. Cory, R. D. Palmiter, and R. L. Brinster, *Nature (London)* **318,** 533 (1985).

6. A. Leder, P. K. Pattengale, A. Kuo, T. A. Stewart, and P. Leder, *Cell (Cambridge, Mass.)* **45**, 485 (1986).
7. V. R. Baichwal and B. Sugden, *Cell (Cambridge, Mass.)* **52**, 787 (1988).
8. T. Dambaugh, C. Beisel, M. Hummel, and E. Kieff, *Proc. Natl. Acad. Sci. U.S.A.* **77**, 2999 (1980).
9. J. Skare and J. L. Strominger, *Proc. Natl. Acad. Sci. U.S.A.* **77**, 3860 (1980).
10. B. Baer, A. Bankier, M. Biggin, P. L. Deninger, P. J. Farrell, T. J. Gibson, G. Hatful, G. Hudson, S. C. Satchwell, C. Seguin, P. S. Tuffnell and B. G. Barrell *Nature (London)* **311**, 207 (1984).
11. J. Skare, J. Farley, J. L. Strominger, K. O. Fresen, M. S. Cho, and H. zurHausen, *J. Virol.* **55**, 286 (1985).
12. S. H. Speck and J. L. Strominger, *Prog. Nucleic Acid Res. Mol. Bio.* **34**, 189 (1987).
13. P. J. Austin, E. Flemington, C. N. Yandava, J. L. Strominger, and S. Speck, *Proc. Natl. Acad. Sci. U.S.A.* **85**, 3678 (1988).
14. G. Laux, M. Perricaudet, and P. J. Farrell, *EMBO J.* **7**, 769 (1988).
15. K. Hennessy, S. Fennwald, M. Hummel, T. Cole, and E. Kieff, *Proc. Natl. Acad. Sci. U.S.A.* **81**, 7207 (1984).
16. B. M. Reedman and G. Klein, *Int. J. Cancer* **11**, 499 (1973).
17. J. L. Yates, N. Warren, D. Reisman, and B. Sugden, *J. Virol.* **81**, 3806 (1984).
18. F. Wang, C. D. Gregory, M. Rowe, A. B. Rickinson, D. Wang, M. Birkenbach, H. Kikutani, T. Kishimoto, and E. Kieff, *Proc. Natl. Acad. Sci. U.S.A.* **84**, 3452 (1987).
19. T. Dambaugh, F. Wang, K. Hennessy, E. Woodland Bushman, A. Rickinson, and E. Kieff, *J. Virol.* **59**, 453 (1986).
20. D. Wang, D. Liebowitz, and E. Kieff, *Cell (Cambridge, Mass.)* **43**, 831 (1985).

8

The Biology of Splicing of Precursors to mRNAs

PHILLIP A. SHARP AND MARIA M. KONARSKA

Center for Cancer Research and Department of Biology
Massachusetts Institute of Technology
Cambridge, Massachusetts

I. Introduction

The excision of introns from precursors of messenger RNA is an essential step in the expression of almost all mammalian genes (*1*). The vast majority of the genes in human cells contain multiple introns. Some of these introns are as short as 75 nucleotides and others as long as several hundred thousand nucleotides. Surprisingly, both the length and sequence of homologous introns in different species vary without apparent restriction. For example, the 5′ proximal introns in the preproinsulin genes of rat and chicken are 0.5 and 3.5 kb in length and show no apparent conserved sequences other than the sequences immediately

THE REGULATION OF PROLIFERATION AND DIFFERENTIATION IN NORMAL AND NEOPLASTIC CELLS

flanking the 5′ and 3′ splice sites (*2*). This latter observation strongly suggests that the sequences within most introns are not informational and that it is likely that these sequences play no role, other than structural, in the physiology of the organism.

Exons are not always spliced together in the same pattern in different cell types. That is, for a small fraction of the genes, different combinations of exons are processed into the mature mRNA in different cell types. For this subset of genes, the splicing process is regulated by cell type-specific factors. An example of regulated splicing is the fibronectin gene. Fibronectins are a family of proteins that are synthesized and secreted by almost all cell types and form part of the extracellular matrix to which cells bind. All the mRNAs for a family of fibronectin proteins are transcribed from a single gene and are related to one another by alernative splicing (*3*). In all cell types, different fibronectin mRNAs are synthesized by splicing to either of two combinations of 5′ and 3′ splice sites. This alternative splicing results in the production of three different proteins which vary in their propensity to form extracellular matrices. In another region of the fibronectin gene, two specific exons are included in mRNAs synthesized in most cell types but excluded in mRNAs synthesized in liver cells. In this case, the splicing pattern is regulated in a cell type-specific fashion (*3*).

II. Biochemical Mechanism of Splicing of mRNA Precursors

A single general mechanism is thought to be responsible for the splicing of all nuclear mRNA precursors in mammalian, plant, and yeast cells (*4*). Analysis of the splicing of radioactive substrate RNA in reactions containing nuclear extracts of either mammalian or yeast cells revealed the generation of *lariat* RNAs. A lariat RNA contains a site where the molecule branches. The excised intervening sequences (IVS) are released as a lariat RNA with the terminal guanosine residue linked through a 2′–5′ phosphodiester bond to an adenosine residue within the intron (Fig. 1). The branch site is typically 20–50 nucleotides upstream of the 3′ splice site. A kinetic intermediate forms during splicing. This intermediate consists of two RNAs, the 5′ exon (E1), and a lariat form of the

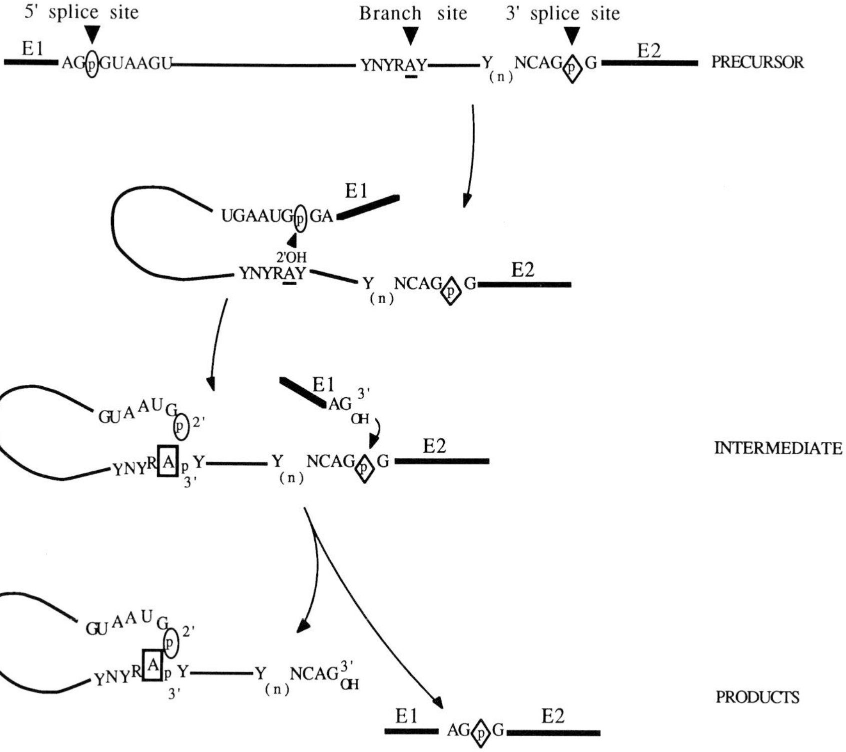

Fig. 1. Splicing mechanism of the mRNA precursor. A prototype precursor RNA is drawn with the intervening sequences or intron sequences spanning from the indicated 5′ and 3′ splice sites. The intervening sequences are flanked by the 5′ exon (E1) and 3′ exon (E2). Consensus sequences are indicated at the splice sites and branch site (Y, pyrimidine; R, purine; and N, any base). The fate of the phosphate moieties at the 5′ and 3′ splice sites during the reaction can be deduced from following the ⓟ and ◇p respectively. The two RNAs of the intermediate are diagrammed in the third line. At the bottom of the figure, the two products of the reaction, the lariat form of the excised intervening sequences and the spliced exons, are shown.

IVS–3′ exon (E2) (Fig. 1). Cleavage at the 5′ splice site and formation of the branch is the first covalent modification of the precursor during splicing. The cleavage and branch formation reactions have not been resolved kinetically and are thought to occur simultaneously. Similarly, the cleavage reaction at the 3′ splice site is thought to occur simultaneously with ligation of the two exons. The phosphate moieties at both the 5′ and the 3′ splice sites are conserved in the products. Thus, both steps may be transesterification reactions, in which a hydroxyl group reacts with a phosphodiester bond, displacing a hydroxyl group while forming a new phosphodiester bond (*5*). This type of reaction is characteristic of RNA-catalyzed processes that do not involve high-energy cofactors such as ATP.

III. Formation of a Spliceosome

The nucleoplasm of animal cells contains five types of snRNP[1] particles, U1, U2, U4, U5, and U6, with an abundance greater than 10^5 copies per cell (*6*). These particles possess between five and nine polypeptides in addition to the RNA component, snRNA. U6 snRNA is distinct from the other snRNAs in that it is transcribed by RNA polymerase III, not RNA polymerase II, and does not have a trimethyl guanine cap structure. This snRNA is frequently found associated with U4 in a U4–6 snRNP. The total complement of snRNAs within animal cells is not known since many low-abundance species may have escaped detection. As an example, another snRNA, U7, was discovered only through a biochemical assay for the processing of the 3′ terminus of a histone mRNA precursor (*7*). This species is present in sea urchin nuclei at a concentration of approximately 5×10^4 copies per cell. Four of these snRNAs share some distinctive properties: a trimethyl cap structure at the 5′ terminus of the snRNA moiety, a common set of core polypeptides recognized by poly- or monoclonal antibodies of

[1] Small nuclear ribonucleoprotein.

the Sm type, and an internal RNA sequence of the type AUUUUUG. This sequence is thought to be responsible for the binding of some of the core polypeptides and, subsequently, further methylation of the cap.

Analysis both *in vivo* and *in vitro* has shown that multiple snRNPs are essential for splicing (8). Thus, it was not surprising to find that the reaction illustrated in Fig. 1 was executed in a complex containing multiple snRNPs. This complex, which contains the lariat RNA characteristic of the intermediate in splicing, is called a *spliceosome* (i.e., splicing body) (*9*). The snRNP composition of the spliceosome has been analyzed by affinity purification using substrate RNA modified by incorporation of biotin and chromatography on streptavidin columns (*10*). This revealed the presence of U2, U4, U5, and U6 snRNAs in the spliceosome. A higher-resolution analysis of splicing complexes was achieved by electrophoresis of reactions in neutral polyacrylamide gels (*11,12*). Three complexes specific for splicing were resolved under these conditions: A, B, and C (Fig. 2) (*13*). Analysis of the kinetics of the reaction showed that the fastest migrating complex, A, appears before complex B, which in turn appeared before the slowest migrating complex, C (13). Thus, complex A could be a precursor to complex B, which in turn could be a precursor to complex C. The lariat RNA characteristic of the intermediate in splicing was found primarily in complex C (13). Furthermore, mutants with altered sequences at the 5′ splice site, which are blocked from proceeding beyond formation of the intermediate RNAs, accumulate in the C complex. This strongly suggests that splicing typically proceeds through sequential formation of the three complexes. A fourth complex, I, was observed after prolonged incubation (*12*). This complex contains the excised intron. It is highly likely that the complex I is rapidly disassociated *in vivo* and the excised intron degraded. This reaction is apparently inefficient *in vitro*. The other products of the splicing reaction, the joined exons, are released as a separate complex which *in vivo* would be rapidly transported to the cytoplasm.

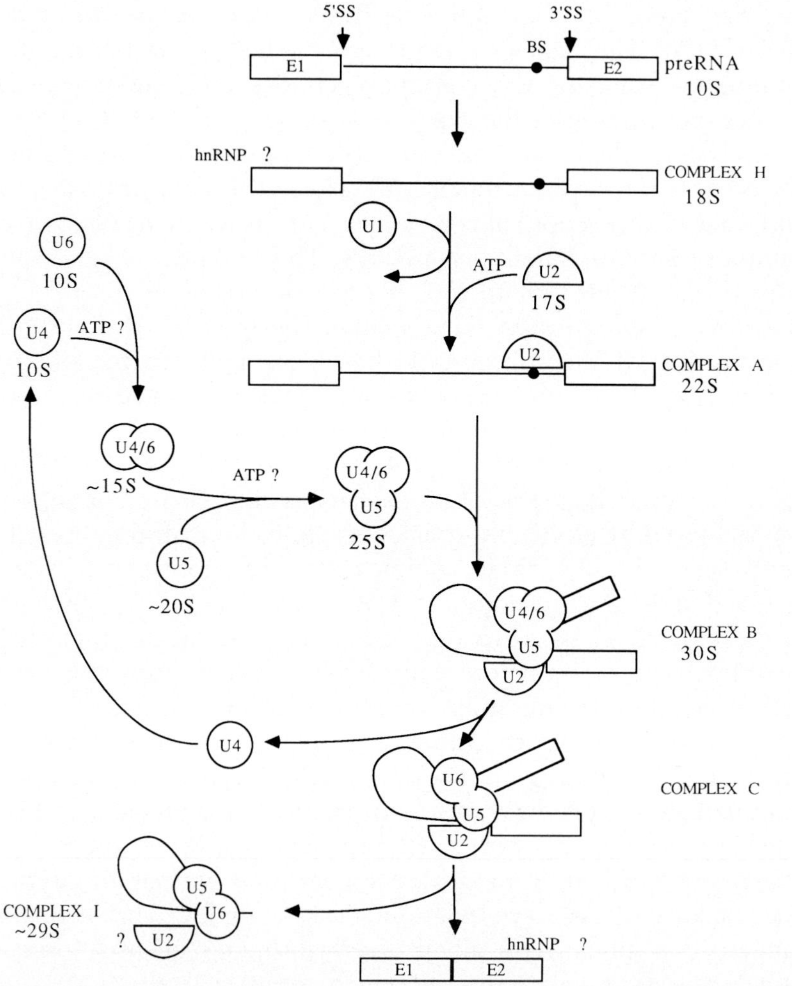

Fig. 2. Schematic representation of involvement of snRNP particles in pre-mRNA splicing. Relative positions and molar ratios of snRNPs in various complexes are arbitrarily illustrated.

IV. snRNP Composition of Splicing Complexes

The snRNP composition of the complexes resolved by electrophoresis in neutral gels can be determined by transferring the contents of the gel in a replica fashion to a membrane and detecting the position of snRNAs by hybridization with specific probes. This analysis reveals that complex A contains only U2 snRNP. Other experiments position the U2 snRNP complex over the branch site and 3′ splice site of the precursor RNA (*11*). Complex B contains U2, U4, U5, and U6 snRNAs. Surprisingly, complex C only contains U2, U5, and U6 snRNAs; thus, U4 snRNA must be selectively released from complex B during conversion to C (*13*). The I complex has the same snRNA composition as complex C. Not surprisingly, the spliced exons are not associated with a snRNP. Both the spliced product and complex H (see Fig. 2), which rapidly and nonspecifically form upon addition of RNA to a reaction, have bound hnRNP proteins (*12*). These basic proteins are important for splicing; however, their specific role in the process is unclear (*14*).

Complexes containing multiple snRNPs, which form in the absence of substrate RNA, can be readily detected in nuclear extracts of mammalian cells by the same electrophoresis and hybridization protocol. The predominant complex in these extracts contains U4, U5, and U6 snRNAs (*12*). We believe this complex directly participates in the conversion of complex A to complex B. Also readily detectable in such extracts are complex U4–U6 and free U4, U5, and U6 snRNP. Incubation of extracts at higher temperatures in the presence of ATP shifts the population of snRNP complexes from the U4–5–6 form to free particles. Thus, their association appears to be quite dynamic. Another, and perhaps more functional, indication of this dynamic nature of the association of snRNPs is the specific release of U4 snRNP during conversion of complex B to complex C. This snRNA is very stable *in vivo* and must participate in the splicing of multiple introns. Its reutilization would then depend upon reassembly of the U4–6 and U4–5–6 complexes. Thus, it is highly likely that snRNP particles are in a constant dynamic association and disassociation in the nucleus of cells.

V. Formation of Pseudospliceosomes

Formation of the U4–5–6 multi-snRNP complex suggests that components in the snRNP particles alone specify recognition of other snRNP particles. These components could either be RNA or protein constituents. It is likely that the structure of the spliceosome is also specified by snRNP–snRNP interactions as contrasted with snRNP–substrate RNA interactions. This is suggested by the formation of a distinct multi-snRNP complex containing all the snRNP constituents of the spliceosome, U2, U4, U5, and U6, in the absence of substrate RNA. As illustrated in Fig. 3, this complex, termed the *pseudospliceosome,* only forms in reactions incubated at high temperatures in high salt concentrations (250 m*M* NH_4 Cl)

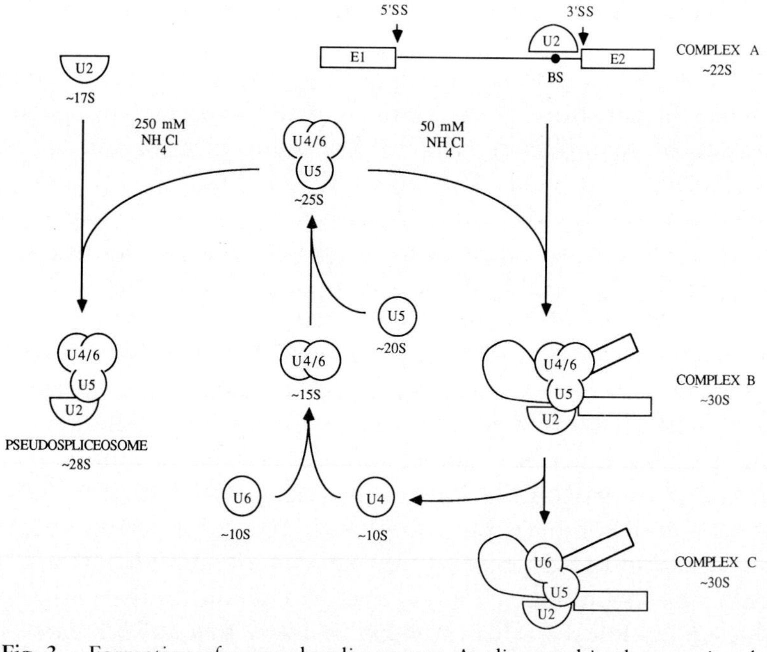

Fig. 3. Formation of a pseudospliceosome. As discussed in the text, incubation of a reaction containing nuclear extract at high ionic strength results in the formation of a multi-snRNP complex with a similar size and composition as the spliceosome. Both the spliceosome and pseudospliceosome are illustrated as forming by association of the U4–5–6 multi-snRNP complex with U2 snRNP. This is assumed to be the case since the vast majority of the three snRNPs in the reaction are in a U4–5–6 complex. However, given the dynamic nature of association of the snRNPs, this is only an assumption.

(*15*). Exogenously added substrate RNA is not incorporated into the complex under these conditions nor is formation of the pseudospliceosome dependent upon the presence of substrate RNA. Thus, the snRNP particles in the spliceosome can specifically recognize one another in the absence of substrate RNA. This snRNP–snRNP interaction probably forms the backbone of the spliceosome.

The generality of formation on a variant complex of snRNP particles during splicing of pre-mRNA was reinforced by analysis of yeast snRNAs and splicing complexes. An analog for each of the mammalian snRNAs U1, U2, U4, U5, and U6 have been identified in the yeast *Saccharomyces cerevisiae*, snR19, snR20, snR14, snR7, and snR6, respectively (*16*). Furthermore, the equivalent analysis of yeast splicing complexes by electrophoresis in native gels yielded a set of complexes homologous to that shown in Fig. 2 (*17,18*). These complexes have the equivalent snRNP composition even to the release of the U4 analog, snR14, during conversion between the two slower-migrating forms. Thus, the snRNP structure of the spliceosome and probably its functions have been conserved over billions of years of evolution.

It is highly likely that during the excision of each intron, a spliceosome of invariant structure is formed encompassing the 5′ and 3′ splice sites. Components, either RNA or protein, in this complex would execute the cleavage and ligation reactions. The specificity for excision of an intron would necessarily reflect specificity in formation of the spliceosome on particular splice sites. Part of the specificity for spliceosome formation is certainly the recognition of local sequences in the vicinity of the 5′ and 3′ splice sites. However, as suggested above, the limited sequence conservation around the two splice sites seems inadequate to specify the formation of a spliceosome on a unique set of splice sites separated by 100,000 nucleotides or more in length. A similar statement can be made concerning regulation of splicing. What controls the particular splice sites utilized in formation of a spliceosome? We do not know the answer to these questions. It is likely that combining *in vitro* and *in vivo* studies with specific mutations in intron sequences will yield the answer.

Acknowledgments

The authors would like to thank their colleagues for contributions to the research described in this article. M.M.K. is a Lucille P. Markey Scholar. This work was supported by a grant from the Lucille P. Markey Charitable Trust (No. 87-19) to M.M.K., by a grant from the National Institutes of Health GM34277, and partially by an NCI core grant PO1-CA14051 to P.A.S.

References

1. R. A. Padgett, P. J. Grabowski, M. M. Konarska, S. Seiler, and P. A. Sharp, *Annu. Rev. Biochem.* **55**, 1119–1150 (1986).
2. F. Perler, A. Efstratiadis, P. Lomedico, W. Gilbert, R. Colodner, and J. Dodgson, *Cell (Cambridge, Mass.)* **20**, 555–566 (1980).
3. R. O. Hynes, *Annu. Rev. Cell Biol.* **1**, 67–90 (1985).
4. P. A. Sharp, *Science* **235**, 766–771 (1987).
5. T. R. Cech and B. L. Boss, *Annu. Rev. Biochem.* **55**, 599–630 (1986).
6. H. Busch, R. Reddy, L. Rothblum, and Y. C. Choi, *Annu. Rev. Biochem.* **51**, 617–654 (1982).
7. M. L. Birnstiel, M. Bussinger, and K. Strub, *Cell (Cambridge, Mass.)* **41**, 349–359 (1985).
8. T. Maniatis and R. Reed, *Nature (London)* **325**, 220–224 (1987).
9. P. J. Grabowski, S. R. Seiler, and P. A. Sharp, *Cell (Cambridge, Mass.)* **42**, 345–353 (1985).
10. P. J. Grabowski and P. A. Sharp, *Science* **233**, 1294–1299 (1986).
11. M. M. Konarska and P. A. Sharp, *Cell (Cambridge, Mass.)* **46**, 845–855 (1986).
12. M. M. Konarska and P. A. Sharp, *Cell (Cambridge, Mass.)* **49**, 763–774 (1987).
13. A. I. Lamond, M. M. Konarska, P. J. Grabowski, and P. A. Sharp, *Proc. Natl. Acad. Sci. U.S.A.* **85**, 411–415 (1988).
14. Y. D. Choi, P. J. Grabowski, P. A. Sharp, and G. Dreyfuss, *Science* **231**, 1534–1539 (1986).
15. M. M. Konarska and P. A. Sharp, unpublished results (1988).
16. P. G. Siliciano, D. A. Brow, H. Roiha, and C. Guthrie, *Cell (Cambridge, Mass.)* **50**, 585–592 (1987).
17. C. W. Pikielny, B. C. Rymond, and M. Rosbach, *Nature (London)* **324**, 341–345 (1986).
18. S.-C. Cheng and J. Abelson, *Genes Dev.* **1**, 1014–1027 (1987).

9

Cloning the 14;19 Translocation Breakpoint in Chronic Lymphocytic Leukemia

TIMOTHY McKEITHAN, MANUEL DIAZ, AND JANET D. ROWLEY
Departments of Medicine and Pathology
University of Chicago
Chicago, Illinois

I. Introduction

Nonrandom chromosome rearrangements are characteristic of neoplasms, particularly the leukemias and lymphomas (Yunis, 1983; LeBeau and Rowley, 1984). In the past few years, investiga-

THE REGULATION OF PROLIFERATION AND DIFFERENTIATION IN NORMAL AND NEOPLASTIC CELLS

tors have identified the genes located at the breakpoints of a few of the recurring chromosome abnormalities. Molecular analysis suggests that the chromosome rearrangement leads to alterations in the expression of these genes or in the properties of the encoded proteins. These alterations are likely to play an integral role in the process of malignant transformation.

A. Genetic Analysis of Translocations in B-Cell Neoplasms

The best studied example among the lymphoid malignancies is that of Burkitt's lymphoma, in which one invariably observes one of three translocations [t(8;14)(q24;q32), t(2;8)(p12;q24), or t(8;22)(q24;q11)]. The consequence of these rearrangements is the juxtaposition of the cellular oncogene *myc,* which is located at band q24 of chromosome 8, and the immunoglobulin heavy chain (*igh*), κ light chain (*igk*), or λ light chain (*igl*) genes, which are located at chromosome bands 14q32, 2p12, and 22q11, respectively. As a result of these translocations, the *myc* gene seems to be transcriptionally deregulated (Croce *et al.,* 1983; Taub *et al.,* 1984; Pelicci *et al.,* 1986). Molecular analysis of the sequences located at the breakpoints of other consistent chromosome rearrangements may identify additional genes of critical importance in the development of neoplasia. The identification of such genes could also provide important new tools to study the factors involved in growth regulation in normal cells.

B-cell neoplasms very frequently show translocations affecting chromosome 14 at band q32, the site of the *igh* locus. Since this locus has been well studied and DNA probes are available, it is possible to identify the DNA sequences involved in the translocation and, by DNA mapping techniques, to identify any gene translocated to a position adjacent to the *igh* locus. By these means, investigators have cloned previously unidentified sequences on chromosomes 11 and 18 involved in the t(11;14) (Tsujimoto *et al.,* 1984a) and t(14;18) (Tsujimoto *et al.,* 1984b) translocations, both of which are found in B-cell lymphomas. In the case of the t(14;18) translocations, this has led to the identification of a gene, *bcl2*, immediately adjacent to the breakpoint on chromosome 18 (Tsujimoto *et al.*, 1985; Bakhshi *et al.*, 1985).

B. Genetic Analysis of the 14;19 Translocation

CLL (chronic lymphocytic leukemia) is the most common form of leukemia in the western world and constitutes over a quarter of the cases of leukemia. The presence of karyotypic abnormalities is associated with poor prognosis (Juliusson *et al.*, 1985; Han *et al.*, 1984). We have reported that the t(14;19)(q32;q13.1) pattern is a recurring translocation in CLL (Ueshima *et al.*, 1985). Of four CLL patients with adequate cytogenetic analysis who had abnormalities involving 14q32, three had a t(14;19) translocation. A similar translocation has been reported by others (Bloomfield *et al.*, 1983; Crossen *et al.*, 1987). Our three patients all had a relatively aggressive form of the disease associated with a shorter than normal survival of only 2–4 years. Although the translocation does not appear to be very frequent, its incidence may be underestimated due to the subtlety of the changes in chromosome appearance resulting from the translocation and to the poor chromosome morphology often found in CLL patients.

DNA from the leukemic cells of two patients was examined using available *igh* probes. The J_H, C_μ, and C_δ gene segments are arranged in that order on chromosome 14, with J_H furthest from the centromere. Other *igh* constant region exons extending from $C_{\gamma 3}$ to $C_{\alpha 1}$ are an unknown distance 3′ (toward the centromere) from these gene segments, and finally, an additional block of exons ($C_{\gamma 2}$ to $C_{\alpha 2}$), separated from the other *igh* constant regions, is located more 3′ (Flanagan and Rabbitts, 1982). Southern blot analysis was performed using probes for J_H and for the μ, γ, and α constant regions.

Complicating the analysis is the fact that deletion of DNA sequences occurs normally during B-cell development, both during the formation of a complete variable exon from V, D, and J segments and in the process of immunoglobulin class switching from synthesis of IgM to synthesis of one of the various classes of IgG, IgA, and IgE. During class switching, deletion of a large stretch of DNA occurs, extending from just 5′ of the $C\mu$ exon to just 5′ of the gene segment to which the switch is being made. Switching results in changes in the restriction fragment size of the latter gene segment and deletion of other gene segments 5′ to it. Thus one has to distinguish the *igh* rearrangements that occur as a result of the normal rearrangement of the gene from those that are associated with the translocation.

C. Case Reports

1. Case 1 (J. L.)

J. L. is a 52-year-old black male diagnosed as having CLL; laboratory analysis revealed an Hb of 14.5 g/dl, white blood cells (WBC) of 100,500/mm^3 with 93% lymphocytes, platelets of 212,000/mm^3, and *IgK* surface immunoglobulin (Ueshima *et al.*, 1985). For 3 months he received cytoxan, vincristine, and prednisone (CVP) therapy and then prednisone prior to the cytogenetic analysis. Chlorambucil was added and then increased because of rising WBC counts. He developed cellulitis of the left arm unresponsive to antibiotics and he died with a septic shock picture 3 years after diagnosis.

At the time of cytogenetic analysis the bone marrow had 6% blasts, 4% prolymphocytes, 11% large lymphocytes, and 79% small lymphocytes with a diagnosis of prolymphocytic transformation of CLL. Five of nine metaphases from the 72-hr culture and 10 of 34 metaphases in the 7dPWM[1] culture were abnormal. The peripheral blood had a total white blood cell count of 61,900/mm^3 with a differential of 26% prolymphocytes, 20% large lymphocytes, and 54% small lymphocytes. Six of 14 metaphases were abnormal in the 7dPWM. The karyotype of the abnormal clone was 45,XY,−9,−14,−17,t(6;?),+der(12)t(12;17)

[1] 7-d pokeweed mitogen.

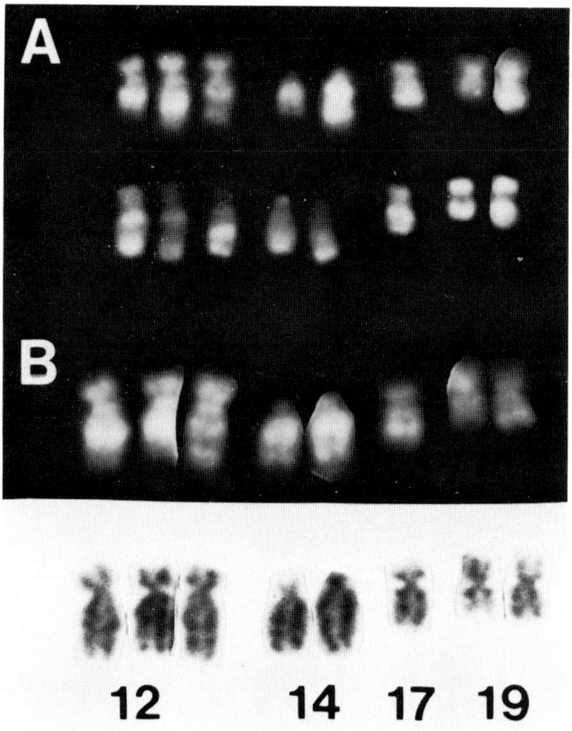

Fig. 1. Partial karyotype of two cells from Case 1. (A) Chromosomes stained with quinacrine mustard (*upper*) or chromomycin A3 and methyl green for R banding (*lower*). (B) Quinacrine mustard and regular Giemsa staining. The third chromosome 12 is abnormal as a result of a t(12;17)(q21.2;q11) translocation. The second 14 and 19 are involved in a translocation [t(14;19)(q32;q13.1)]; additional material of unknown origin is added to the 19q+ chromosome (second 19).

(q21.2;q11),+der(14)t(14;19)(q32;q13.1),t(19;?)(q13.1;?) (Fig. 1 and 2). One cell had the t(14;19) as well as three normal chromosomes 12. The major clone showed further rearrangement of one chromosome 12 with material translocated to 12q21 which appeared to come from chromosome 17 on Q and R banding. A 19q+ was also present; the material could not be definitively identified but may have been derived from 17q.

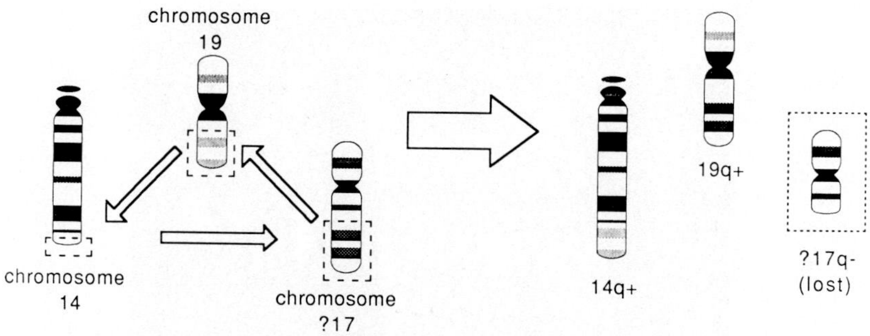

Fig. 2. Schematic diagram illustrating our interpretation of the chromosome rearrangements that have occurred in cases 1 and 2. The long arm of No. 19 is translocated to chromosome 14(14q+); the long arm of chromosome 14 is translocated to a chromosome, possibly 17, which is missing from the cell; additional material is translocated to chromosome 19(19q+) which may come from chromosome 17.

2. *Case 2 (D. B.)*

D. B. was a 58-year-old white male with a diagnosis of CLL (Ueshima *et al.*, 1985). Laboratory analysis showed an Hb of 15.9 g/dl, WBC of 22,000/mm^3 with 60% lymphocytes, and platelets of 120,000/mm^3. He required no therapy for 2 years, when he received chlorambucil and then prednisone, followed 10 months later by CVP, because of decreasing Hb/Hct and platelets. Five months later, he again received chlorambucil and prednisone, and he was on this medication for 5 months when a specimen was sent for cytogenetic analysis. The diagnosis of prolymphocytoid transformation of CLL was made. His disease progressed with falling Hb/Hct and platelets and rising WBC counts and finally agranulocytosis and disseminated herpes zoster; he expired 4 years after the initial diagnosis.

At the time of analysis, he had 14% prolymphocytes, 5% large lymphocytes, and 72% small lymphocytes in the bone marrow with a diagnosis of prolymphocytic transformation of CLL. Seven of nine metaphases from the 24-hr culture were abnormal. The peripheral blood had a total white blood cell count of 489,000/mm^3 (20% prolymphocytes, 15% large lymphocytes, 56% small lymphocytes, 4% neutrophils, 1% eosinophils, 1% basophils, 3% monocytes). The abnormal clone, 45,XY,−14,−17,−19,+der (14)t (14;19) (q32;q13), + der(19)t (?17;19) (?q21;q13),t (2;14) (p13;q32),was present in the 24-hr and 4-day cytochalasin B cultures(three of four metaphases). This case had a complex rearrangement involving chromosomes 14, 17, and 19. The exact location of the breakpoint in chromosome 19 was very difficult to determine using both G and Q banding.

II. Results

Both cases of t(14;19) were studied extensively by Southern blot analysis and showed multiple rearrangements and deletions with *igh*. In both cases it was possible to construct a model consistent with the Southern blot results; the models suggested that in each patient the t(14;19) rearrangement involved a break in one of the C_α segments of *igh*. The rearranged C_α bands were cloned and were shown to contain the chromosome breakpoint juctions. In one case, there is evidence for a transcribed gene on chromosome 19 adjacent to the breakpoint junction.

A. Analysis of Case 1 (J. L.)

In J. L., the t(14;19) translocation was part of the three-way chromosome rearrangement. Most of 19q was translocated to 14q32, the site of *igh;* the distal portion of 14q was lost; and additional material, possibly from 17q, was translocated to 19q (Fig. 2). Southern blot analysis showed deletion of both C_μ alleles, one J_H, both $C_{\alpha 1}$ s, and all but one of the C_γ alleles (McKeithan *et al.*, 1987) (Fig. 3). The remaining J_H, C_γ, and one of the two $C_{\alpha 2}$s were rearranged. Additional probes revealed that the remaining J_H is associated with the remaining C_γ. The simplest interpretation of the results from Southern blot analysis is as follows: VDJ joining and class switching to $C_{\gamma 4}$ occurred on the normal chromosome 14 with deletion of all constant region gene segments other than $C_{\alpha 2}$

and $C_{\gamma 4}$, which was rearranged; the translocation involved the other chromosome 14 with a break close to $C_{\alpha 2}$, with loss of all genetic elements 5′ of the break due to the absence of the distal portion of 14q from the karyotype (Fig. 4).

Clones containing rearranged C_{α} sequences were isolated from a λ library made using a complete *Bgl*II digest. Detailed mapping of one clone revealed that the 3′ portion contains the $C_{\alpha 2}$ exon. The map of the major part of the clone does not resemble that of any mapped region of *igh*, suggesting that the clone contains a chromosome breakpoint junction. A 0.5-kb *Bam*HI fragment free of repetitive sequences was subcloned ($p_{\alpha .5B}$) and mapped to chromosome 19 using a panel of 31 different human–mouse somatic cell hybrids obtained from the laboratory of Thomas Shows. A cell hybrid provided by Anthony Carrano in which 19q was the only human DNA present was positive with this probe,

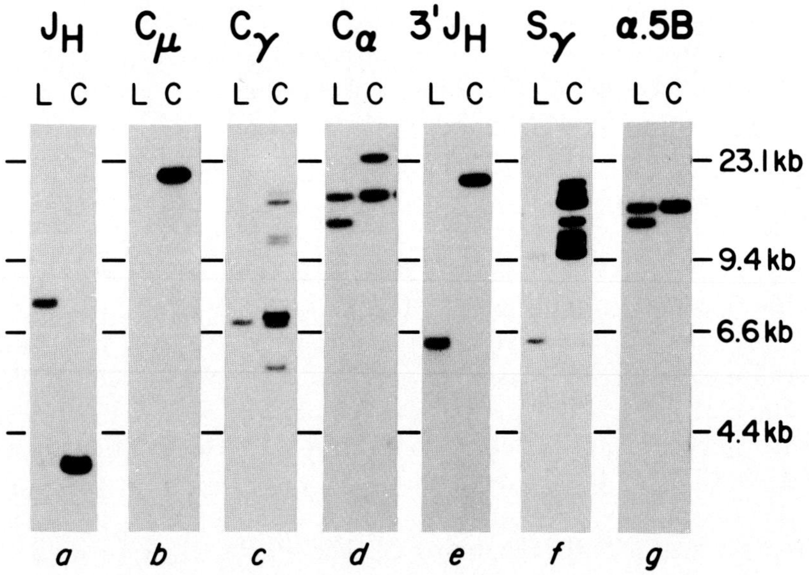

Fig. 3. Southern blot hybridization of *Bgl*II digests by using probes of *igh*. Lanes: L,DNA from patient J. L.; C, DNA from control (placenta). (a–g) Autoradiograms of a blot hybridized successively to the probes as illustrated.

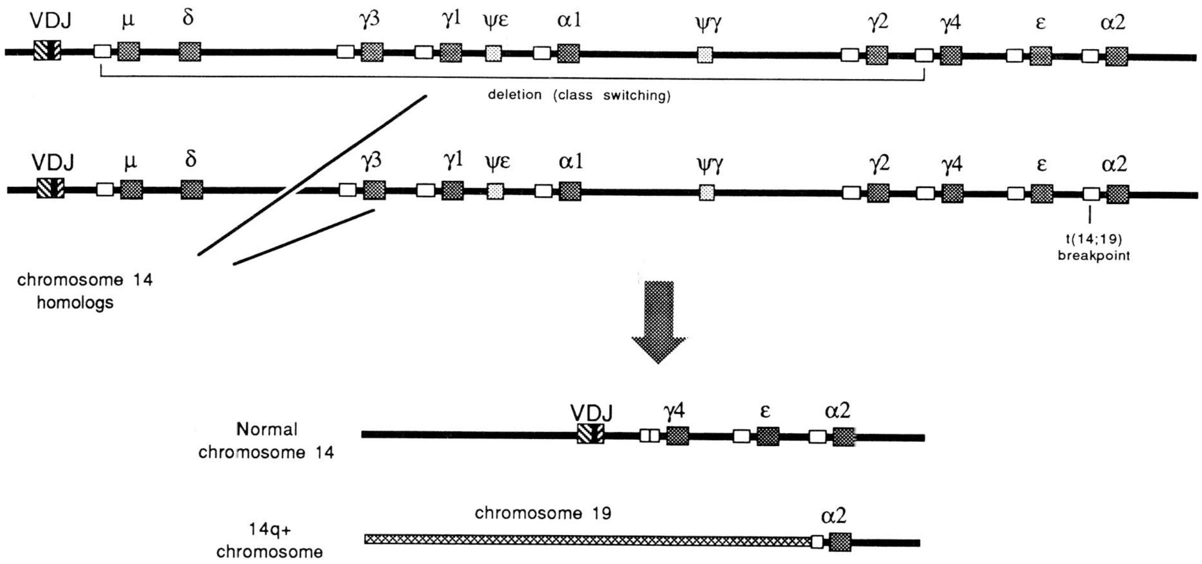

Fig. 4. A schematic diagram of the rearrangements which have probably occurred in case J. L. VDJ joining is assumed to have occurred initially. A normal class switch to $C_{\gamma 4}$ has resulted in deletion of most of the constant region gene segments. The translocation breakpoint is within the switch region of $C_{\alpha 2}$. All sequences 5′ of this are missing due to loss from the karyotype of the derivative of the third chromosome involved in the translocation. While the model shown is the simplest consistent with experimental results, other more complex models cannot be excluded.

confirming that in the t(14;19) translocation, the break in 19 occurs in the q arm.

Normal sequences from chromosome 19 were cloned using the $p_{\alpha.5B}$ probe. The map of this region matches the portion of the map of the rearranged C_α clone assumed to be from chromosome 19; thus, no DNA rearrangements were found other than that due to the translocation. Comparison of the normal maps of 14 and 19 sequences indicates that the break occurred within the switch region of $C_{\alpha 2}$. Initial hybridizations to Northern blots failed to show evidence for transcription from this region of chromosome 19.

B. Analysis of Case 2 (D. B.)—Evidence for an Involved Gene on Chromosome 19

The t(14;19) rearrangement in this patient is also part of a three-way translocation, in this case also probably involving chromosome 17; as in case 1, there is loss of the chromosome derived from the third partner. In this patient, a second translocation involves the other chromosome 14 and a chromosome 2. Several examples of t(2;14) translocations have been reported. The breakpoint at both chromosomes 14 occurs at q32, the site of the *igh* locus.

Southern blot analysis of DNA from the leukemic cells from patient D.B. revealed multiple rearrangements of components of *igh*. One copy each of C_μ and J_H has been deleted. Molecular cloning showed that the remaining J_H and C_μ are adjacent, with deletion of much of the intervening immunoglobulin class switch region (S_μ). Analysis with a probe for the C_αs shows rearrangement of one $C_{\alpha 1}$ and deletion of the other copy; both copies of $C_{\alpha 2}$ appear to remain and are unrearranged. The rearranged $C_{\alpha 1}$ band was cloned from a *Bgl*II library. This clone was mapped and a fragment free of repetitive sequences was subcloned ($p_{\alpha l.4P}$) and, using somatic cell hybrids, was assigned to chromosome 19.

Most of the human genome is depleted in CG dinucleotides;

however, despite representing only about 1% of the total genome, regions called CpG islands containing a much higher frequency of this dinucleotide are found in or adjacent to all "housekeeping" genes studied, as well as many tissue-specific genes (Bird, 1986). Thus, we examined the clones for the presence of restriction sites for enzymes recognizing sequences which include one or more CGs. The subclone $p_{\alpha 1.4P}$ contains three *Sma*I sites (one CG) and two *Sac*II sites (two CGs). It has been estimated that 47% of *Sma*I sites and 74% of *Sac*II sites are found within CpG islands (Brown and Bird, 1986). The large number of sites in $p_{\alpha 1.4P}$ suggests that it is part of a CpG island and may be part of a gene or adjacent to a gene.

Hybridization of the probe $p_{\alpha 1.4P}$ to a Northern blot showed significant hybridization to a band of 2.3 kb. The transcript is found in the RNA of MOLT16, a T-cell acute lymphoblastic leukemia cell line, and, to a lesser extent, in BV173, a cell line from a patient with chronic myelogenous leukemia in blast crisis. These results show that probe $p_{\alpha 1.4P}$, adjacent to the t(14;19) breakpoint, contains part of a transcribed gene. Further analysis of this region of chromosome 19 shows that the breakpoint in cells from D.B. is about 19 kb from that in cells from J. L.

III. Discussion

A. Translocations in B-Cell Neoplasms

The cloning of the t(14;19) translocation described in this paper is important from both the biological and the clinical points of view. When one considers the number of recurring translocations that have been described in leukemia, lymphoma, and in the sarcomas, relatively few translocation breakpoint junctions have been cloned. For the B-cell neoplasms, these include the three translocations in Burkitt lymphoma and B-cell ALL (acute lymphocytic leukemia) which involve *myc* on chromosome 8 and one of the immunoglobulin genes, either the heavy chain gene at 14q32, or the κ or λ light chain genes at 2p12 or 22q11, respectively. Other translocations in neoplastic B cells that have been cloned include the t(11;14) rearrangement that occurs in CLL and in lymphomas and the t(14;18) pattern, which is a very common translocation in follicular lymphomas. These translocations, including those associated with the endemic form of Burkitt's lymphoma, involve the *igh* gene, with breakpoints primarily in the J region. The DNA probe cloned on chromosome 11 has been called *BCL1* (B-cell leukemia 1) and at the present time no mRNA has been identified with this probe (Tsujimoto *et al.,* 1984a). Therefore, the nature of the gene at this breakpoint has not been determined. In contrast, the gene on chromosome 18, termed *BCL2,* has been analyzed extensively. The breakpoint junction was cloned in three different laboratories (Tsujimoto *et al.,* 1984b; Bakhshi *et al.,* 1985; Cleary and Sklar, 1985). In fact, there are at least two junctions; the more frequent is termed the major breakpoint region (mbr) and is estimated to account for about 60–70% of all t(14;18) translocations. Although the function of *BCL2* has not been established, the translocation results in the switching off of the normal *BCL2* gene and the expression of mRNAs of abnormal size from the translocated *BCL2* (Tsujimoto and Croce, 1986; Cleary *et al.,* 1986). The abnormal mRNAs code for a normal BCL2 protein because the fusion of the *BCL2* and *igh* genes occurs in the 3′ untranslated portion of *BCL2*. The t(14;19), therefore, is the sixth translocation breakpoint junction to be cloned in B-cell tumors.

B. Structural Rearrangements in T-Cell Neoplasms

Seven translocation junctions have been cloned in malignant T cells involving genes for both the α and β chains of the T-cell receptors. These include t(8;14)(q24;q11) translocations in T-cell CLL or ALL (Shima *et al.,* 1986; McKeithan *et al.,* 1986; Mathieu-Mahul *et al.,* 1985; Erikson *et al.,* 1986) and two different t(11;14) rearrangements, one with a break in 11p13 (Erikson *et al.,* 1985; Lewis *et al.,* 1985) and the other with a break in 11p15 (Boehm *et al.*, 1988). The inv(14)(q11q32) inversion in T-cell CLL or ALL is especially complex because it can involve breaks in 14q11 in the J region of *tcra* and in 14q32 either in the V region of the *igh* gene (Baer *et al.*, 1985; Denny *et al.*, 1986) or 3′ (centromeric) of *igh* (Baer *et al.*, 1987). Each of these rearrangements involves the joining of the 5′ end of the J segment (either J_α or, more rarely, J_δ), with loss of the recombination signal at the breakpoint in the translocation or inversion. This implies that the V–J recombinase is involved in these rearrangements and that there has been inappropriate recombination of the *tcra* gene with a gene on another chromosome that contains the appropriate recognition signals even though it is not a "lymphoid" gene (Finger *et al.*, 1986; Baer *et al.*, 1987). In the t(8;14) rearrangement the break in chromosome 8 involves *myc;* the genes on chromosome 11 in the two different t(11;14) translocations have not been identified. Several translocations involving the β chain of the T-cell receptor (*tcrb*) localized to 7q34 (LeBeau *et al.*, 1985) have been studied. The t(7;9)(q34;q34) rearrangement was shown to involve *tcrb* (Reynolds *et al.*, 1987) but not *abl* (Westbrook *et al.*, 1987). Another translocation t(7;19)(q34;p13) joins *tcrb* just 5′ of $J_{\beta 1.1}$ with an unknown gene on 19, called *lyl1* (Cleary *et al.*, 1988).

C. Functional Role of Translocations

Certain paradigms have been developed based primarily on the translocations involving *myc* and either *igh* or *tcra*. In these translocations, a cell specific gene is moved adjacent to *myc*, which presumably is involved in the regulation of cell proliferation (Rowley, 1984; Klein and Klein, 1986). The specificity of the translocations is directly related to the cell-specific gene, which, in each of these examples, is a very highly active gene in the appropriate cell. Thus, the immunoglobulin genes are involved in Burkitt's lymphoma, and *tcra* is involved in T-cell leukemia. Applying this paradigm to the t(14;19) translocation, we predict that *BCL3* on chromosome 19 has a role in the regulation of cell proliferation and may be a growth factor, a growth factor receptor, or a nuclear binding protein.

From the clinical point of view, cloning of *BCL3* may have an impact in several areas. First, we have found that our patients with a t(14;19) rearrangement have had a particularly aggressive form of CLL with prolymphocytoid transformation and a survival of $2\frac{1}{2}$, 3, and 4 years from diagnosis. They also required the early institution of therapy. The two patients described by Bloomfield *et al.* (1983) also transformed from CLL to small lymphocytic or diffuse mixed small and large cell lymphoma. Thus, it may be possible to screen DNA from CLL patients to identify those who have a *BCL3* rearrangement and who may have a more aggressive form of CLL. In the patients we have studied, we do not know whether the t(14;19) translocation was present at the development of the disease or whether it occurred during the evolution to a more malignant phenotype. However, based on the morphology of the bone marrow and peripheral blood samples at diagnosis, these patients had features suggesting an aggressive disease from the onset. One of the patients reported by Bloomfield was studied prior to therapy and had the translocation as the only abnormality. It will be important to use this probe in prospective as well as in retrospective studies to determine the frequency of this rearrangement and the clinical implications of *BCL*3 rearrangement. Moreover, because cytogenetic analysis of CLL samples is hampered by a low mitotic index, studies of these patients are performed only infrequently. The use of DNA analysis with a panel of cloned probes should provide an important resource for correlating clinical and pathologic findings with the DNA analysis to identify useful prognostic parameters.

Finally, information about the protein that is produced, its function in normal cells, and its alteration in malignant cells will be useful. Although quite some time will be required for successful completion of this research, it may be possible to use the change in this protein to identify leukemic cells and to target therapy specifically to the affected cells while sparing the normal cells.

Thus, chromosome translocations provide a unique opportunity to identify genes that are unequivocally involved in the process of malignant transformation. Cloning the genes at these breakpoints will lead to the discovery of many new genes whose involvement in cancer had not been previously recognized. In the future, an understanding of the function of these genes may lead to new therapeutic strategies which are more effective and less toxic than present regimes.

Acknowledgments

These studies were supported by grants NIH 3-P01-CA19266 and DE-FG02-86ER60408 from the Department of Energy to J. D. R., Institutional Grant IN-41Z from the American Cancer Society, and Grant 87-19 from the American Cancer Society (Illinois Division) to T.W.M. T.W.M. was supported as a Special Fellow of the Leukemia Society of America.

References

Baer, R., Chen, K. C., Smith, S. D., and Rabbitts, T. H. (1985). Fusion of an immunoglobulin variable gene and a T-cell receptor constant gene in the chromosome 14 inversion associated with T-cell tumors. *Cell (Cambridge, Mass.)* **43**, 705–713.

Baer, R., Heppel, A., Taylor, A. M. R., Rabbitts, P. H., Boullier, B., and Rabbitts, T. H. (1987). The breakpoint of an inversion of chromosome 14 in a T-cell leukemia: Sequences downstream of the immunoglobulin heavy chain locus are implicated in tumorigenesis. *Proc. Natl. Acad. Sci. U.S.A.* **84**, 9069–9073.

Bakhshi, A., Jensen, J. P., Goldman, P., Wright, J. J., McBride, O. W., Epstein, A. L., and Korsmeyer, S. J. (1985). Cloning the chromosomal breakpoint of t(14;18) human lymphomas: Clustering around J_H on chromosome 14 and near transcriptional unit on 18. *Cell (Cambridge, Mass.)* **41**, 899–906.

Bird, A. P. (1986). CpG-rich islands and the function of DNA methylation. *Nature (London)* **321**, 209–213.

Bloomfield, C. D., Arthur, D. C., Frizzera, G., Levine, E. G., Peterson, B. A., and Gaji-Peczalska, K. J. (1983). Nonrandom chromosome abnormalities in lymphoma. *Cancer Res.* **43**, 2975–2984.

Boehm, T., Baer, R., Lavenir, I., Forster, A., Waters, J. J., Nacheva, E., and Rabbits, T. H. (1988). The mechanism of chromosomal translocation t(11;14) involving the T-cell receptor Cδ locus on human chromosome 14q11 and a transcribed region of chromosome 11p15. *EMBO J.* **7,** 385–394.

Brown, W. R. A., and Bird, A. P. (1986). Long-range restriction site mapping of mammalian genomic DNA. *Nature (London)* **322,** 477-481.

Cleary, M. L., and Sklar, J. (1985). Nucleotide sequence of a t(14;18) chromosomal breakpoint in follicular lymphoma and demonstration of a breakpoint cluster region near a transcriptionally active locus on chromosome 18. *Proc. Natl. Acad. Sci. U.S.A.* **82,** 7439–7443.

Cleary, M. L., Smith, S. D., and Sklar, J. (1986). Cloning and structural analysis of cDNAs for bcl-2 and a hybrid bcl-2/immunoglobulin transcript resulting from the t(14;18) translocation. *Cell (Cambridge, Mass.)* **47,** 19–28.

Cleary, M. L., Mellentin, J. D., Spies, J., and Smith, S. D. (1988). Chromosomal translocation involving the β T-cell receptor gene in acute leukemia. *J. Exp. Med.* **167,** 682–687.

Croce, C. M., Thierfeld, W., Erikson, J., Nishikura, K., Finan, J., Lenoir, G. M., and Nowell, P. C. (1983). Transcriptional activation of an unrearranged and untranslocated c-*myc* by translocation of a C locus in Burkitt lymphoma. *Proc. Natl. Acad. Sci. U.S.A.* **80,** 6922–6926.

Crossen, P. E., Goodwin, J. M., Heaton, D. C., and Tully, S. M. (1987). Chromosome abnormalities in chronic lymphocytic leukemia revealed by Cytochalasin B and Epstein-Barr Virus. *Cancer Genet. Cytogenet.* **28,** 93–100.

Denny, C. T., Yoshikai, Y., Mak, T. W., Smith, S. D., Hollis, G. F., and Kirsch, I. R. (1986). A chromosome 14 inversion in a T-cell lymphoma is caused by site-specific recombination between immunoglobulin and T-cell receptor loci. *Nature (London)* **320,** 549–551.

Erikson, J., Williams, D. L., Finan, J., Nowell; P. C., and Croce, C. M. (1985). Locus of the α-chain of the T-cell receptor is split by chromosome translocation in T-cell leukemias. *Science* **229,** 784–786.

Erikson, J., Finger, L., Sun, L., Ar-Rushdi, A., Nishikura, K., Minowada, J., Finan, J., Emanuel, B. S., Nowell, P. C., and Croce, C. M. (1986). Deregulation of c-*myc* by translocation of the α-locus of the T-cell receptor in T-cell leukemias. *Science* **232,** 884–886.

Finger, L. R., Harvey, R. C., Moore, R. C. A., Showe, L. C., and Croce, C. M. (1986). A common mechanism of chromosomal translocation in T and B-cell neoplasia. *Science* **234,** 982–985.

Flanagan, J. G., and Rabbitts, T. H. (1982). Arrangement of human immunoglobulin heavy chain constant region genes implies evolutionary duplication of a segment containing γ, ε and α genes. *Nature (London)* **300,** 709–713.

Han, T., Ozer, H., Sadamori, N., Emrich, L., Gomez, G. A., Henderson, E. S., Blom, M. L., and Sandberg, A. A. (1984). Prognostic importance of cytogenetic abnormalities in chronic lymphocytic leukemia. *N. Engl. J. Med.* **310,** 288.

Juliusson, G., Robert K.-H., Ost, A., Friberg, K., Biberfeld, P., Nilsson, B., Zech, L., and Gahrton, G. (1985). Prognostic information from cytogenetic analysis in chronic B-lymphocytic leukemia and leukemic immunocytoma. *Blood,* **65,** 134–141.

Klein, G., and Klein, E. (1986). Conditional tumorigenicity of activated oncogenes. *Cancer Res.* **46,** 3211–3224.

LeBeau, M. M., and Rowley, J. D. (1984). Recurring chromosomal abnormalities in leukemia and lymphoma. *Cancer Surv.* **3,** 371–394.

LeBeau, M. M., Diaz, M. O., Rowley, J. D., and Mak, T. W. (1985). Chromosomal localization of the human T-cell receptor beta-chain genes. *Cell (Cambridge, Mass.)* **41,** 335.

Lewis, W. H., Michalopoulos, E. E., Williams, D. L., Minden, M. D., and Mak, T. W. (1985). Breakpoints in the human T-cell antigen receptor α-chain locus in two T-cell leukemia patients with chromosomal translocations. *Nature (London)* **317,** 544–546.

McKeithan, T. W., Shima, E. A., LeBeau, M. M., Minowada, J., Rowley, J. D., and Diaz, M. O. (1986). Molecular cloning of the breakpoint junction of a human chromosomal 8;14 translocation involving the T-cell receptor alpha-chain gene and sequences on the 3′ side of c-*myc*. *Proc. Natl. Acad. Sci. U.S.A.* **83,** 6636–6640.

McKeithan, T. W., Rowley, J. D., Shows, T. B., and Diaz, M. O. (1987). Cloning of the chromosome translocation breakpoint junction of the t(14;19) in chronic lymphocytic leukemia. *Proc. Natl. Acad. Sci. U.S.A.* **84,** 9257–9260.

Mathieu-Mahul, D., Caubet, J. F., Bernheim, A., Mauchauffe, M., Palmer, E., Berger, R., and Larsen, C.-J. (1985). Molecular cloning of DNA fragment from human chromosome 14(14q11) involved in T-cell malignancies. *EMBO J.* **4,** 3427–3433.

Pelicci, P.-G., Knowles, D. M., Magrath, I., and Dalla-Favera, R. (1986). Chromosomal breakpoints and structural alterations of the c-*myc* locus differ in endemic and sporadic forms of Burkitt lymphoma. *Proc. Natl. Acad. Sci. U.S.A.* **83,** 2984–2988.

Reynolds, T. C., Smith, S. D., and Sklar, J. (1987). Analysis of DNA surrounding the breakpoints of chromosomal translocations involving the β T-cell receptor gene in human lymphoblastic neoplasms. *Cell (Cambridge, Mass.)* **50,** 107–117.

Rowley, J. D. (1984). Biological implications of consistent chromosome rearrangements in leukemia and lymphoma. *Cancer Res.* **44,** 3159–3168.

Shima, E. A., LeBeau, M. M., McKeithan, T. W., Minowada, J., Showe, L. C., Mak, T. W., Minden, M. D., Rowley, J. D., and Diaz, M. O. (1986). Gene encoding of the alpha chain of the T-cell receptor is moved immediately downstream of *c-myc* in a chromosomal 8;14 translocation in a cell line from a human T-cell leukemia. *Proc. Natl. Acad. Sci. U.S.A.* **83,** 3439–3443.

Taub, R., Moulding, C., Battey, J., Murphy, W., Vasicek, T., Lenoir, G. M., and Leder, P. (1984). Activation and somatic mutation of the translocated c-*myc* gene in Burkitt lymphoma cells. *Cell (Cambridge, Mass.)* **36,** 339–348.

Tsujimoto, Y., and Croce, C. M. (1986). Analysis of the structure, transcripts, and protein products of *bcl-2,* the gene involved in human follicular lymphoma. *Proc. Natl. Acad. Sci. U.S.A.* **83,** 5214–5218.

Tsujimoto, Y., Yunis, J., Onorato-Showe, L., Erikson, J., Nowell, P. C., and Croce, C. M. (1984a). Molecular cloning of the chromosomal breakpoint of B-cell lymphomas and leukemias with the t(11;14) chromosome translocation. *Science* **224,** 1403–1406.

Tsujimoto, Y., Yunis, J., Nowell, P. C., and Croce, C. M. (1984b). Cloning of the chromosome breakpoint of neoplastic B-cells with the t(14;18) chromosome translocation. *Science* **226,** 1097–1099.

Tsujimoto, Y., Cossman, J., Jaffe, E., and Croce, C. M. (1985). Involvement of the bcl-2 gene in human follicular lymphoma. *Science* **228,** 1440–1443.

Ueshima, Y., Bird, M. L., Vardiman, J., and Rowley, J. D. (1985). A 14;19 translocation in B-cell chronic lymphocytic leukemia: A new recurring chromosome aberration. *Int. J. Cancer* **36,** 287–290.

Westbrook, C. A., Rubin, C. M., LeBeau, M. M., Kaminer, L. S., Diaz, M. O., Smith, S. D., and Rowley, J. D. (1987). Molecular analysis of *tcrb* and *abl* in a human T-cell leukemia cell line (SUP-T3) with a chromosomal 7;9 translocation. *Proc. Natl. Acad. Sci. U.S.A.* **84,** 251–255.

Yunis, J. J. (1983). The chromosomal basis of neoplasia. *Science* **221,** 227–235.

10

The *mos* and *met* Oncogenes: Transformation and Reverse Genetics*

GEORGE F. VANDE WOUDE, MARY GONZATTI-HACES, ANAND IYER, MORAG PARK, JOSEPH R. TESTA, MARIANNE OSKARSSON, RICHARD S. PAULES, FRIEDRICH PROPST, AND NORIYUKI SAGATA

BRI-Basic Research Program
Frederick, Maryland

I. Introduction

Oncogenes have been identified based on their capacity to induce tumors in animals or cause morphological transformation *in vitro*. They are diverse in function and their products have

* Research sponsored by the National Cancer Institute, DHHS, under Contract NO. NO1-CO-74101 with Bionetics Research, Inc. The contents of this publication do not necessarily reflect the views or policies of the Department of Health and Human Services, nor does mention of trade names, commercial products, or organizations imply endorsement by the United States Government.

been found to include secreted growth factors, integral transmembrane and nonintegral membrane-associated protein kinases, and nuclear proteins with nucleic acid binding activity. They are derived from normal cellular genes (proto-oncogenes) and there are several mechanisms by which they can be activated as oncogenes, but in general the changes alter either normal gene regulation or functional properties of the gene product. The *mos* oncogene is an example of the former, while the specific *met* oncogene activation is of the latter type.

Oncogenes were first identified in acute transforming retroviruses and *mos* is one of many that have been discovered this way. Genomic DNA transfer–transfection assays have served to identify additional oncogenes and the *tpr–met* oncogene was identified by use of this assay. The fundamental questions we wish to ask are, what alterations are critical to the activation of the oncogene, and how do these products contribute to expression of the transformed phenotype. The products of oncogenes presumably display at least a subset of the activities of the normal proto-oncogene. Most importantly, therefore, we would like to determine the normal function of proto-oncogene products. For a variety of reasons this often requires that proto-oncogene homologs be characterized in animal model systems and in genetically or developmentally well-characterized organisms. The study of an oncogene, then, can require the use of reverse genetics, in which a segment of DNA with transforming properties is used to isolate the normal homolog from a variety of species. This permits characterization of the transforming potential of its products by providing the opportunity to make defined mutations in the normal gene and to test the effect of these mutations when reintroduced into the cells or animal of origin.

We have focused primarily on the characterization of the *met* and *mos* oncogenes. What follows is an overview of our attempts to understand their normal function.

II. Results and Discussion

A. The *met* Oncogene

The *met* oncogene was activated in a cell line derived from a human osteogenic sarcoma (HOS) after prolonged *in vitro* exposure to the carcinogen *N*-methylnitronitrosoguanidine (Rhim *et al.*, 1975). We have shown that *met* is activated via a DNA rearrangement (Park *et al.*, 1986a) and is a member of the tyrosine kinase family of growth factor receptors (Park *et al.*, 1987; Gonzatti-Haces *et al.*, 1988).

The properties of the *met* oncogene locus are summarized in Figs. 1 and 2. We have determined that *met* was activated via a DNA rearrangement that fused sequences on human chromosome 1 to chromosome 7 (Park *et al.*, 1986a). We refer to the sequences on chromosome 1 as *tpr* (translocated promoter region) (Park *et al.*, 1986a) (Fig. 1). These sequences provide the upstream portion of the oncogene while *met*, on chromosome 7, provides the downstream or the 3′ portion of the oncogene (Fig. 1). The name *tpr–met* oncogene is reserved for this specific rearrangement. A novel 5.0-kb fusion transcript is expressed by the *tpr–met* oncogene (Park *et al.*, 1986a) (Fig. 2). Probes from the *tpr* locus recognize a 10.0-kb RNA transcript in all human cells analyzed, while a 9.0-kb *met* proto-oncogene transcript is detected in human fibroblast and epithelial cancer cell lines with the *met* probes (Fig. 2) and this transcript is 3′ coterminal with the 5.0-kb *tpr–met* oncogene transcript.

To compare the oncogene locus to the proto-oncogene and to begin to characterize the latter, we isolated cDNA corresponding to the 5.0-kb and 9.0-kb transcripts, respectively (Park *et al.*, 1987) (Fig. 3). The nucleotide sequence of the proto-oncogene cDNA reveals an open reading frame of 4224 nucleotides, which codes for a protein 1408 residues in length. This protein has structural features in common with the family of tyrosine kinase–growth factor receptors. Thus, a putative extracellular domain 926 amino acid residues in length is preceded by an amino-

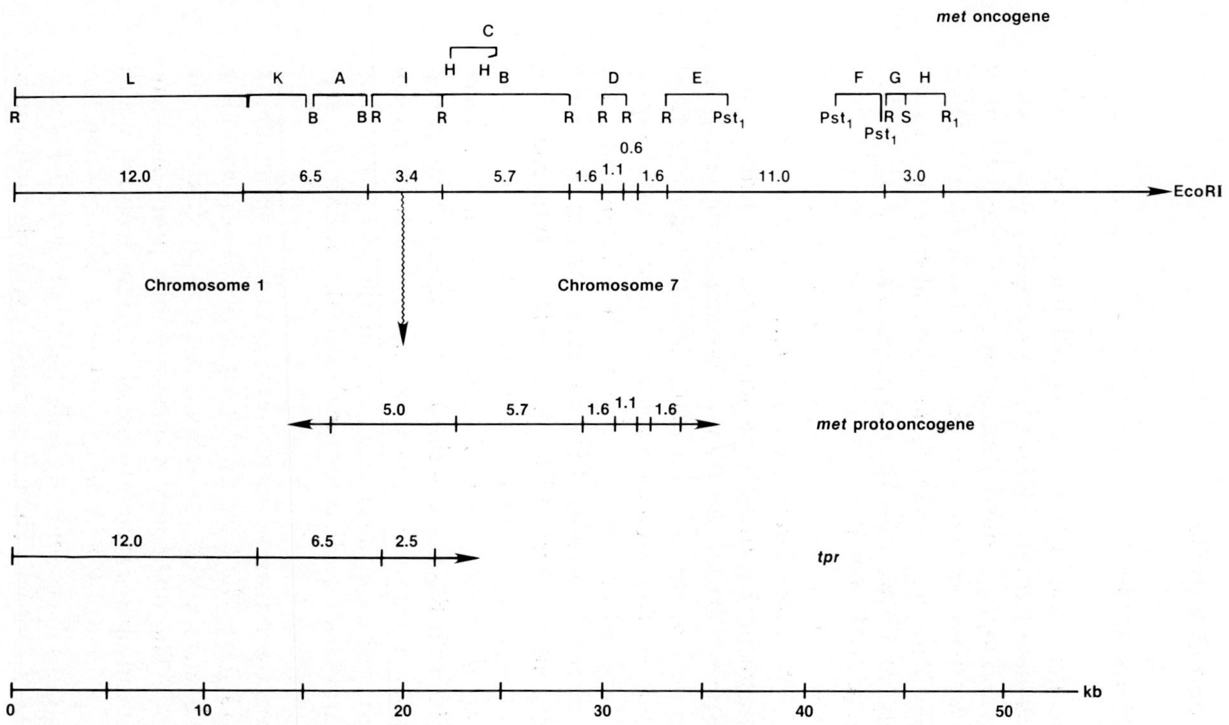

Fig. 1. Partial restriction map of the *tpr–met* oncogene locus, *met* proto-oncogene locus, and the *tpr* locus. A 5.0-kb *Eco*RI restriction fragment is the normal upstream restriction fragment of the *met* proto-oncogene on chromosome 7, whereas a 2.5-kb *Eco*RI restriction fragment is the normal downstream restriction fragment of the *tpr* locus on chromosome 1. Recombination has occurred within these fragments to generate the rearranged fragment I (3.4 kb) of the *tpr–met* oncogene.

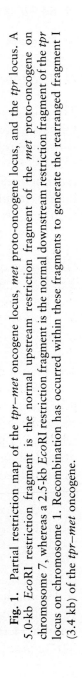

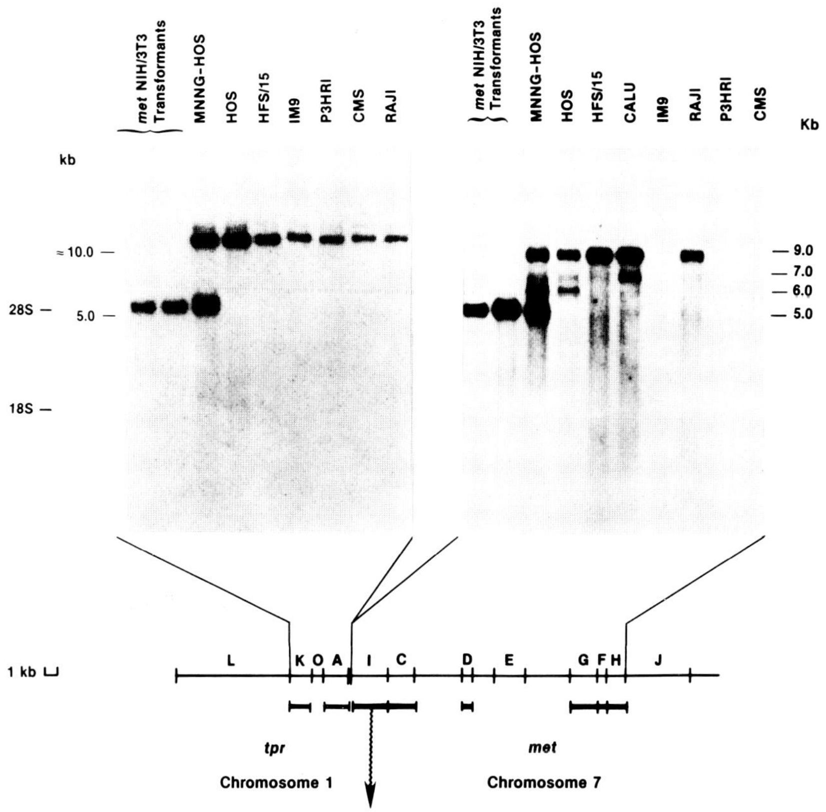

Fig. 2. Northern blot analysis of *met*-related RNA. Fragments L, M, K, O, A, I, C, D, E, P, Q, G, F, H, and J were used as probes against RNA prepared from human cell lines. Each probe was hybridized to 10 μg of poly (A)+ RNA prepared from *met* NIH-3T3 transformants, MNNG-HOS cells, HOS cells, HFS/15 (human foreskin fibroblast cells), IM9 (human B-cell line), p3HR1, RAJI (human Burkitt's lymphoma cell lines), CMS (human myeloid-derived cell line), and CALU-1 (human lung carcinoma cell line). Fragments showing positive hybridization are underlined; K or A hybridize to the 5.0-kb *met* oncogene RNA in *met* NIH-3T3 transformants and MNNG-HOS cells, and to an RNA of ~10 kb in every human cell line tested. However, fragments I, C, D, G, F, and H, which also hybridize to the 5.0-kb *met* oncogene RNA, hybridize predominantly to a 9.0-kb RNA species in fibroblast and epithelial cells. With the exception of RAJI cells, these probes do not hybridize with RNA prepared from hemopoietic-derived cell lines. An additional RNA of 7.0 kb is detected in CALU-1 cells and two RNAs of 7.0 kb and 6.0 kb are detected in the HOS cell line and its derivatives.

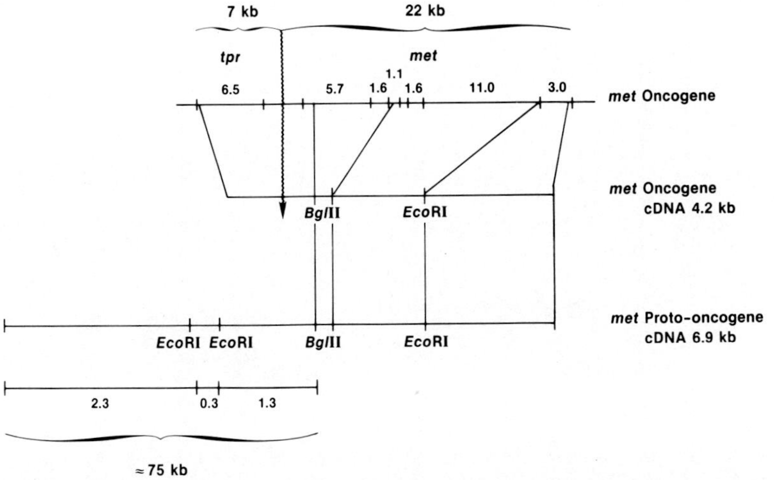

Fig. 3. Schematic representation of *met* cDNA. cDNA for the human *tpr–met* oncogene and the *met* proto-oncogene are aligned with respect to the *tpr–met* oncogene genomic locus.

terminal consensus signal peptide region of 24 amino acids (aa) and followed by a candidate hydrophobic membrane-spanning region 23 aa in length. A putative intracellular carboxy-terminal domain 435 aa long shows 30–40% homology to the tyrosine kinase domains of other *src* family members such as v-*abl* and human insulin receptor (HIR) (Park *et al.*, 1987). A comparison of the putative *met* proto-oncogene protein structure with known tyrosine kinase growth factor receptors (Fig. 4) shows that the external domain displays a cysteine-rich array and the tyrosine kinase domain is not interrupted. The structure is more similar to EGF receptor (EGFR) and HIR than to PDGF or CSF-1 (Park *et al.*, 1987). A comparison of the *tpr–met* oncogene with the *met* proto-oncogene reveals that only the kinase domain (beginning 54 aa downstream from the transmembrane domain) is retained in the activated *tpr–met* oncogene cDNA (Park *et al.*, 1987) (Fig. 4). Thus, *met* may be activated by decapitation in a fashion similar to EGFR in v-*erbB* (Hayman *et al.*, 1983; Privalsky and Bishop, 1984; Ullrich *et al.*, 1984). However, amino-terminal *tpr* coding

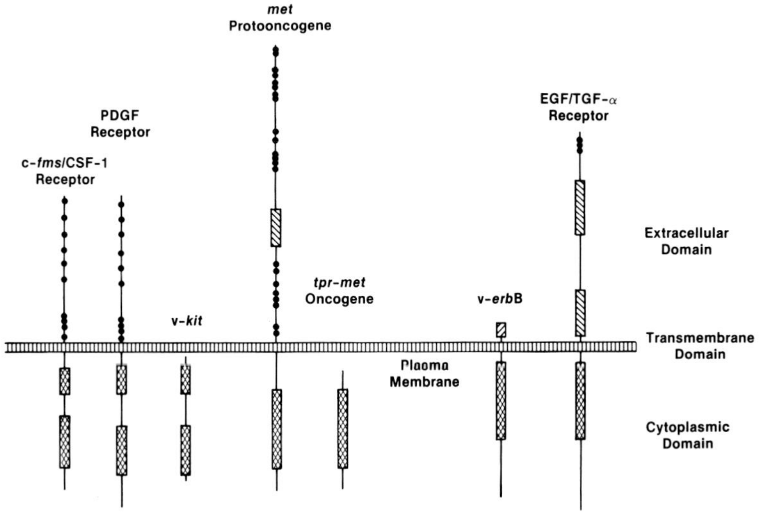

Fig. 4. Schematic comparison of *met* proto-oncogene with other cell surface receptors. Cysteine-rich domains are shown in hatched boxes; other cysteine residues in the extracellular domain are represented as solid circles. Tyrosine kinase domains are cross-hatched boxes. Receptors for CSF-1, colony stimulating factor 1; PDGF, platelet-derived growth factor; EGF/TGF-α, epidermal growth factors.

sequences (Chan *et al.*, 1987) are present in *tpr–met* and may be required for transformation. Similar to other examples of activated oncogenes such as the tropomyosin–*trk* oncogene (Martin-Zanca *et al.*, 1986) or the *bcr–abl* gene in chronic myeloid leukemia (Collins and Groudine, 1983; Heisterkamp *et al.*, 1983; Konopka *et al.*, 1984; Ben-Neriah *et al.*, 1986), the presence of *tpr* may misdirect, or alter, functional properties or targets of the proto-oncogene.

To characterize the *tpr–met* and proto-oncogene products, we prepared antibody to the carboxy-terminal 28 residues of the open reading frame (Park *et al.*, 1986b; Gonzatti-Haces *et al.*, 1988). This peptide antibody detects a 65-kDa polypeptide expressed by the *tpr–met* oncogene (Fig. 5) as well as 110-, 140-, and 160-kDa products expressed by the *met* proto-oncogene locus (Fig. 5) (Park

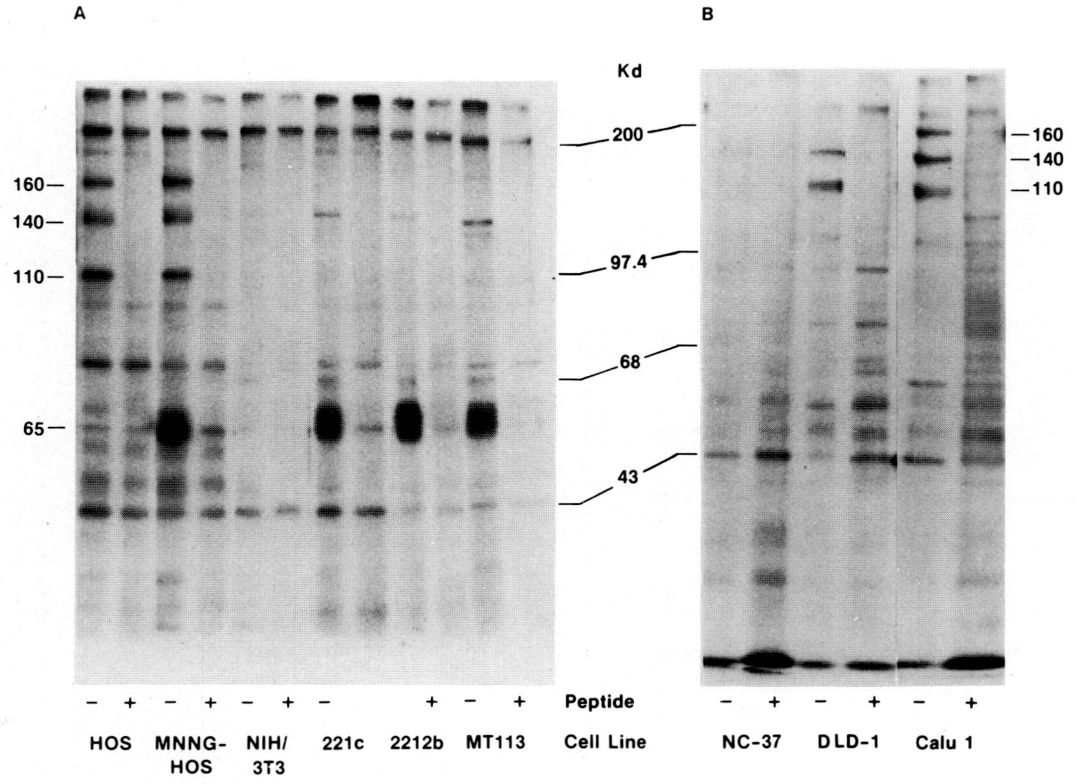

A
B
Kd
160
140
110
65
200
97.4
68
43
160
140
110
– + – + – + – + – + Peptide
HOS
MNNG-HOS
NIH/3T3
221c
2212b
MT113
Cell Line
– + – + – +
NC-37
DLD-1
Calu 1

et al., 1986b; Gonzatti-Haces *et al.*, 1988). The p140*met* proto-oncogene appears to be the mature product and is the major polypeptide detected on the surface of ^{125}I-labeled cells (Gonzatti-Haces *et al.*, 1988). We have also shown that both p65*tpr–met* and p140*met* have tyrosine kinase activity in *in vitro* immune complex assays and will phosphorylate enolase as a substrate (Gonzatti-Haces *et al.*, 1988). *In vivo* we have only found p65*tpr–met* to be phosphorylated on tyrosine and serine while p140*met* from several cell lines labeled *in vivo* with ^{32}P is phosphorylated on serine and threonine residues (Gonzatti-Haces *et al.*, 1988).

In addition to characterizing the human *met* proto-oncogene locus, we wished to develop an animal model system. For this purpose, we isolated cDNA prepared from a C3H mouse embryo fibroblast cDNA library homologous to the human *met* proto-oncogene cDNA (Fig. 6). We estimate by heteroduplex analysis that the 5′ 4.2 kb containing the *met* open reading frame of mouse and human are >90% homologous (A. Iyer, unpublished results) and sequence analyses of several regions of the mouse *met* cDNA also show high conservation of nucleotide sequence with the human. We have found, as with the EGFR (Di Fiore *et al.*, 1987; Hudziak *et al.*, 1987) that the mouse *met* cDNA under the transcription control of an SV40 promoter–enhancer transforms NIH/3T3 cells with an efficiency of 200 foci/pmole (Fig. 6). The transformed cells contain multiple copies of the construct and

---

Fig. 5. Immunoprecipitation of *tpr–met* oncogene and *met* proto-oncogene proteins with anti-C-28 antibody. (A) Two human cell lines, HOS and MNNG-HOS, and four mouse cell lines, including control NIH/3T3 cells and *met* NIH/3T3 transformants, 221c, 2212b, and MT113 were labeled with [^{35}S]methionine for 2 hr. Labeled cell extracts were immunoprecipitated with anti-C-28 *met* antibody pre-incubated in the presence (+) or in the absence (−) of excess peptide. (B) Similar analyses were carried out with a Burkitt's lymphoma B cell line, NC-37, and with two human tumor epithelial cell lines, DLD-1 and CALU-1. Immunoprecipitated proteins were resolved by 8% SDS-polyacrylamide electrophoresis and visualized by fluorography.

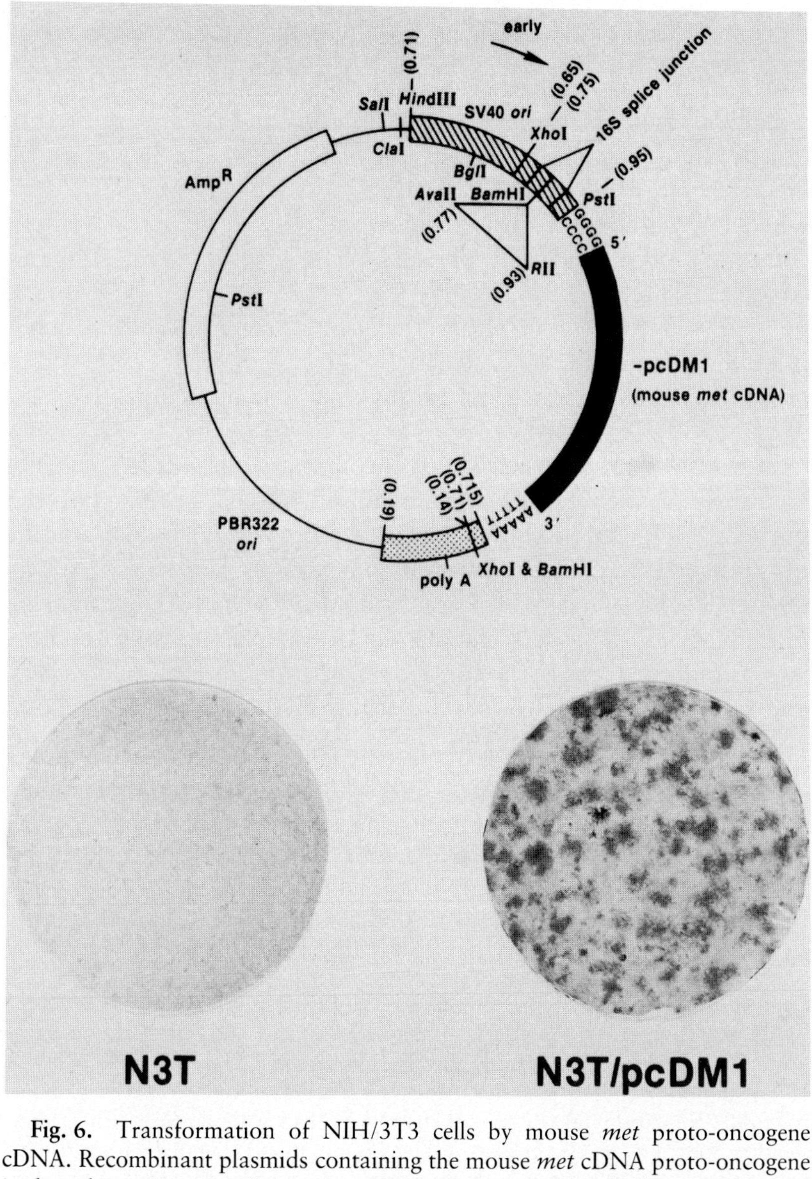

Fig. 6. Transformation of NIH/3T3 cells by mouse *met* proto-oncogene cDNA. Recombinant plasmids containing the mouse *met* cDNA proto-oncogene in the eukaryotic expression vector pcDM1 (*top*) were introduced into NIH/3T3 cells by cotransfection with a selectable drug-resistance marker (G418). The cultures were incubated for 10–14 days in medium supplemented with G418 and 5% fetal bovine serum. The cultures were then fixed and stained with crystal violet.

express relatively high levels of RNA. We are currently attempting to identify the product. We have also examined RNA prepared from adult mouse tissue for *met* expression. We find high levels of expression in skin, lungs, and kidney, but further studies indicate that *met* is expressed at low levels in a wide variety of tissues (A. Iyer, unpublished results). These analyses should help to provide clues for identifying the *met* ligand much as the correlation was made between the v-*fms* oncogene and CSF-1 receptor (Sherr *et al.*, 1985). We can then proceed to elucidate the *met* signal transduction pathway.

B. The *mos* Oncogene

The *mos* oncogene was originally identified as the transforming gene (v-*mos*) of the acute transforming retrovirus Moloney murine sarcoma virus (Mo-MSV) (Frankel and Fischinger, 1976; Vande Woude *et al.*, 1980). This virus causes fibrosarcomas in mice and transforms fibroblasts in culture (Moloney, 1966; Aaronson and Rowe, 1970). The v-*mos* region contains an open reading frame that encodes a 37-kDa *env–mos* fusion protein which has been detected in the cytoplasm of cells acutely infected or transformed by Mo-MSV (Papkoff *et al.*, 1983). The *mos* oncogene is a serine kinase member of the *src* family of kinases (Maxwell and Arlinghaus, 1985). The single-copy cellular homolog of v-*mos* was shown to be colinear with the viral gene (Oskarsson *et al.*, 1980; Jones *et al.*, 1980) and sequence analysis revealed that with the exception of additional amino-terminal amino acids, the v-*mos* of the HT-1 strain of Mo-MSV was identical in coding sequence to the mouse proto-oncogene (Seth and Vande Woude, 1985). This identity was also reflected in similar transforming efficiencies in DNA transfection assays of NIH/3T3 cells when the mouse *mos* oncogene and proto-oncogene were activated by a Mo-MSV long terminal repeat (LTR) (Blair *et al.*, 1980, 1981). In contrast, the human c-*mos* gene (Watson *et al.*, 1982), which shows 77% homology to mouse c-*mos*, transforms NIH/3T3 cells 100-fold

```
c-mos^hu   1   M P S P L A L R P Y L R S E F S P S V D A R P C S S P S E L P - - - A K L L L G - A T L P R A P R L
c-mos^mk               L A L R P Y   P S   F     S V   A   P     S   S E L P - - - A   L L     - A T P     A P R
c-mos^mu               L S L C R Y   P R   L     S V   S   S     I   L V A P R K A G   L F     - T T P     A P R
c-mos^rt               L I L C R Y   P R   L     T V   S   S     S   L V A S R - A G   - F     - A T P     A P G
c-mos^ch               I P F N S F   P L   L     S A   L   P     S   V V I P G K D G   A F     G T P S     T R R

c-mos^hu  47   P R R L A W C S I D W E Q V C L L Q R L G A G G F G S V Y K A T Y R G V P V A I K Q V N K C T K N R
c-mos^mk       P R         C S       E Q V     L Q R     A         S             H     P     I       N   C T K N R
c-mos^mu       P R         F S       E Q V     M H R     S         S             H     P     I       N   C T K N R
c-mos^rt       S R         C F       G Q V     L H R     S         S             H     P     I       N   C T K D L
c-mos^ch       P P         C S       D R L     L Q P     S         A             H     T     V       K   S S K N R

c-mos^hu  97   L A S R R S F W A E L N V A R L R H D N I V R V V A A S T R T P A G S N S L G T I I M E F G G N V T
c-mos^mk       L     R R S             V       R       I V               R T   A G S                   F G
c-mos^mu       R     Q R S             I       R       I V               R T   E D S                   F G
c-mos^rt       R     Q R N             I       H       I I               R T   E G S                   F G
c-mos^ch       L     R Q S             V       Q       V V               C A   A S Q                   Y V

c-mos^hu 147   L H Q V I Y G A A G H - P E G D A G E P H C R T G G Q L S L G K C L K Y S L D V V N G L L F L H S Q
c-mos^mk           Q         A A S H - P E G D A G E P H C S T G G P   T L G K C L K     L   V   N     L
c-mos^mu           Q         A T R S - P E - - - - - P L S C - R E Q   S L G K C L K     L   V   N     L
c-mos^rt           Q         A T R S - P E - - - - - P L S C - R E Q   S L G K C L K     L   I   N     L
c-mos^ch           H         - T R D A W R Q G E E E E G G C G R K A   S M A E A V C     C   I   T     A

c-mos^hu 196   S I V H L D L K P A N I L I S E Q D V C K I S D F G C S E K L E D L L C F Q T P S Y P L G G T Y T H
c-mos^mk       S   V                       S   Q D V       S           E K   E D L L C F Q T P L Y P L
c-mos^mu       S   L                       S   Q D V       S           Q K   Q D L R C R Q A S P H H I
c-mos^rt       S   L                       S   K D V       S           Q K   Q D L R C R - P S L H H I
c-mos^ch       G   V                       T   H G A       G           Q R   E E G L S Q S H H V C Q Q

c-mos^hu 246   R A P E L L K G E G V T P K A D I Y S F A I T L W Q M T T K Q A P Y S G E R Q H I L Y A V V A Y D L
c-mos^mk       R       L         G V   P               A           M T T K Q A     S     R   H I L             D
c-mos^mu       Q       I         I A   P               G           M T T R E V     S     P   Y V Q             N
c-mos^rt       Q       L         I A   P               G           M T T R E V     S     P   H V Q             N
c-mos^ch       R       L         R V   A               A           I V M R E Q     L     R   Y V L             N

c-mos^hu 296   R P S L S A A V F E D S L P G Q R L G D V I Q R C W R P S A A Q R P S A R L L L V D L T S L K A E L G
c-mos^mk           S L S A   V   Q D   L P   Q R   G D V I R R     R P S A A Q   P S   R P   L V D   T S L K A E F G
c-mos^mu           S L A G   V   T A   L T   K T   Q N I I Q S     E A R A L Q   P G   E L   Q R D   K A F R G A L G
c-mos^rt           H W Q -   V   T A   L T   K T   Q N N V Q S     E A R A L Q   P G   E L   Q K D   K A F R G A L G
c-mos^ch           P L A A   I   H E   A V   Q R   R S I I S C     K A D V E E   L S   A Q   L P S   R A L K E N L -
```

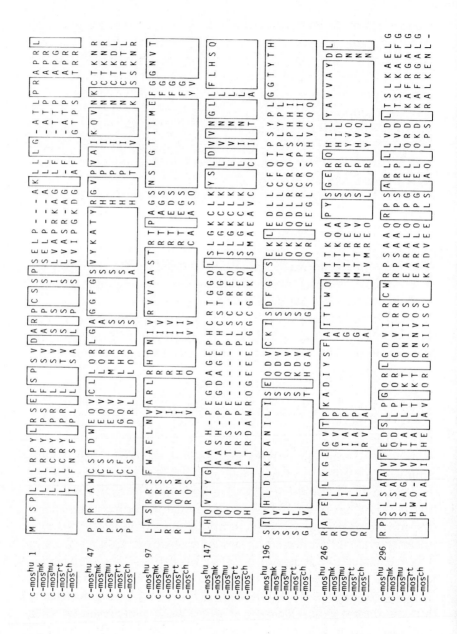

less efficiently than v-*mos* or mouse c-*mos* when linked to an Mo-MSV LTR (Blair *et al.*, 1986).

Valuable information about proto-oncogenes and their physiological function has been obtained by studying their expression in normal cells. However, expression of the c-*mos* gene had not been detected in normal mouse cells or tissues (Gattoni *et al.*, 1982; Muller *et al.*, 1982; Maguire *et al.*, 1983; Goyette *et al.*, 1984). The conservation of the c-*mos* open reading frame in rodents, primates, and chickens (Fig. 7) (Van Beveren *et al.*, 1981; Watson *et al.*, 1982; van der Hoorn and Firzlaff, 1984; Schmidt *et al.*, 1988; R. Paules *et al.*, 1988). strongly indicated that the gene must function during some portion of the animal life cycle, and other lines of evidence (Papkoff *et al.*, 1982; Rechavi *et al.*, 1982; Cohen *et al.*, 1983; Gattoni-Celli *et al.*, 1983; Wood *et al.*, 1983a,b) suggested to us that the c-*mos* proto-oncogene may be expressed at very low levels in normal tissues.

Using a sensitive S1 nuclease assay, we showed that *mos* was expressed at low levels in a number of adult tissues and in mouse embryos (Propst and Vande Woude, 1985). These studies also revealed that the proto-oncogene was expressed at highest levels in gonadal tissues (Propst and Vande Woude, 1985). These results and additional characterization of the *mos* transcripts indicated that the size varied in a tissue-specific fashion due to variations in the 5′ untranslated leader preceding the *mos* open reading frame (Propst *et al.*, 1987). Also, all transcripts characterized thus far do not appear to be processed (spliced) and only contain the *mos* coding region (Fig. 8). We and others have demonstrated that the mouse *mos* gonadal transcripts are expressed in male and female

---

Fig. 7. Comparison of the predicted amino acid sequences of the c-*mos* proto-oncogene of human (Watson *et al.*, 1982), monkey (R. Paules, *et al.*, 1988), mouse (Van Beveren *et al.*, 1981), rat (van der Hoorn and Firzlaff, 1984), and chicken (Schmidt *et al.*, 1988). Amino acids common in all five species are boxed and shown only for c-mos^{hu}. Numbering is on left and corresponds to the amino acids of c-mos^{hu}. Gaps (dashes) have been introduced to align sequences maximally.

germ cells (Propst *et al.*, 1987; Goldman *et al.*, 1987; Mutter and Wolgemuth, 1987; Propst *et al.*, 1988a; Keshet *et al.*, 1988). In the testes the transcript (1.7 kb in length) is developmentally regulated, appears concomitantly with the onset of spermatogenesis, and is found predominantly in postmeiotic spermatids. *mos* transcripts are also expressed in primate gonadal tissues (Fig. 8). While the primate ovarian *mos* transcript is similar to the mouse, the primate testis-specific transcript is markedly different and begins within the *mos* open reading frame (Fig. 8). A protein product expressed from the RNA could initiate at a conserved internal *mos* AUG (Fig. 8) but would lack the ATP binding

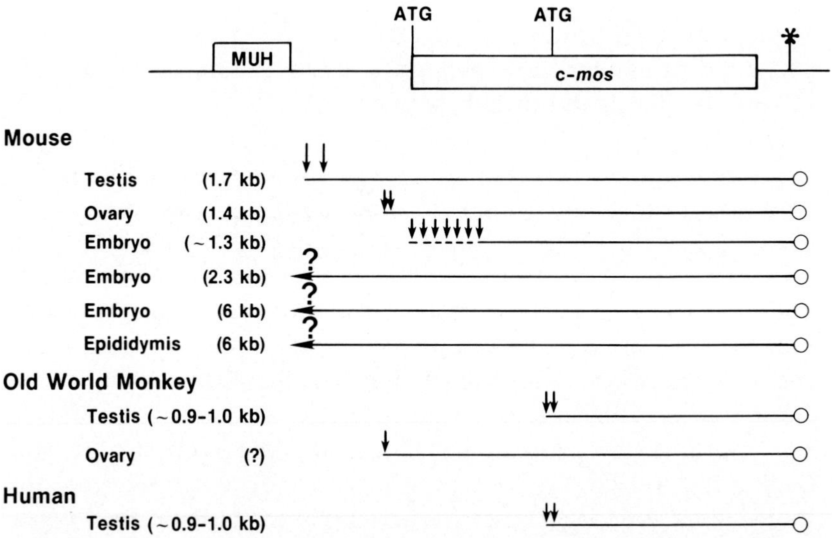

Fig. 8. The genomic *mos* locus and *mos* transcripts detected in various tissues of mouse and human. Vertical arrows and open circles mark the putative 5′ and 3′ ends, respectively, of the various transcripts as indicated by the results of the S1 nuclease protection assay. MUH designates a previously defined region of sequence homology between mouse and human *mos* loci upstream from the *mos* open reading frame, which is indicated by the open box. The asterisk marks a consensus polyadenylation signal 110 bp downstream from c-*mos*. All transcripts, with the exception of the 1.3-kb transcripts in embryonic RNA, contain the entire *mos* open reading frame.

domain (Paules *et al.*, 1988; Propst *et al.*, 1988b). Similar transcription regulation has been shown for an internal transcript of the E2 gene product of bovine papilloma virus (Lambert *et al.*, 1987) encoding a trans-acting negative regulator which shows slight homology to a region in the C terminus of *mos* (Danos and Yaniv, 1984). We have also determined that avian *mos* is expressed in chicken and quail ovaries and testes (Schmidt *et al.*, 1988). Two transcripts are detected in ovaries (3.3 and 1.4 kb) (Fig. 9), while only a 1.4-kb *mos* RNA transcript is detected in testes. In contrast to the mammalian *mos* transcripts, the 3.3-kb avian transcript appears to extend further downstream (Schmidt *et al.*, 1988). Thus, in all animal species examined thus far, *mos* transcripts appear to be expressed at the highest levels in ovaries.

In mouse ovaries the *mos* transcript is present at high levels during oocyte maturation (Goldman *et al.*, 1987; Mutter and Wolgemuth, 1987; Keshet *et al.*, 1988) and ovulation but is not detectable by *in situ* hybridization after fertilization and before fusion of the pronuclei (Keshet *et al.*, 1988). This indicates that *mos* transcripts in the mouse are part of the maternal mRNA pool which diminishes after fertilization and prior to the onset of embryonic transcription.

We have recently isolated the *Xenopus laevis mos* genomic homolog (Sagata *et al.*, 1988). It shows only 54% homology with either the chicken, mouse, or human *mos* locus, but as in the other species, it appears to be a single uninterrupted coding exon. Characterization of the RNA transcript with the *Xenopus* genomic locus does not reveal evidence for processing and the tissue-specific expression pattern is similar to mouse (i.e., brain, testes, and ovaries express *mos* transcripts). In contrast to other species, only one major transcript is detected in these tissues and the transcript is considerably longer (~2 kb) downstream from the *mos* open reading frame. In *Xenopus* high levels of *mos* expression are detected in the earliest stages of egg development (I and II) and this level appears to remain constant throughout oogenesis and oocyte maturation as well as through late blastula (data not shown). This is in contrast to its absence after fertilization in the

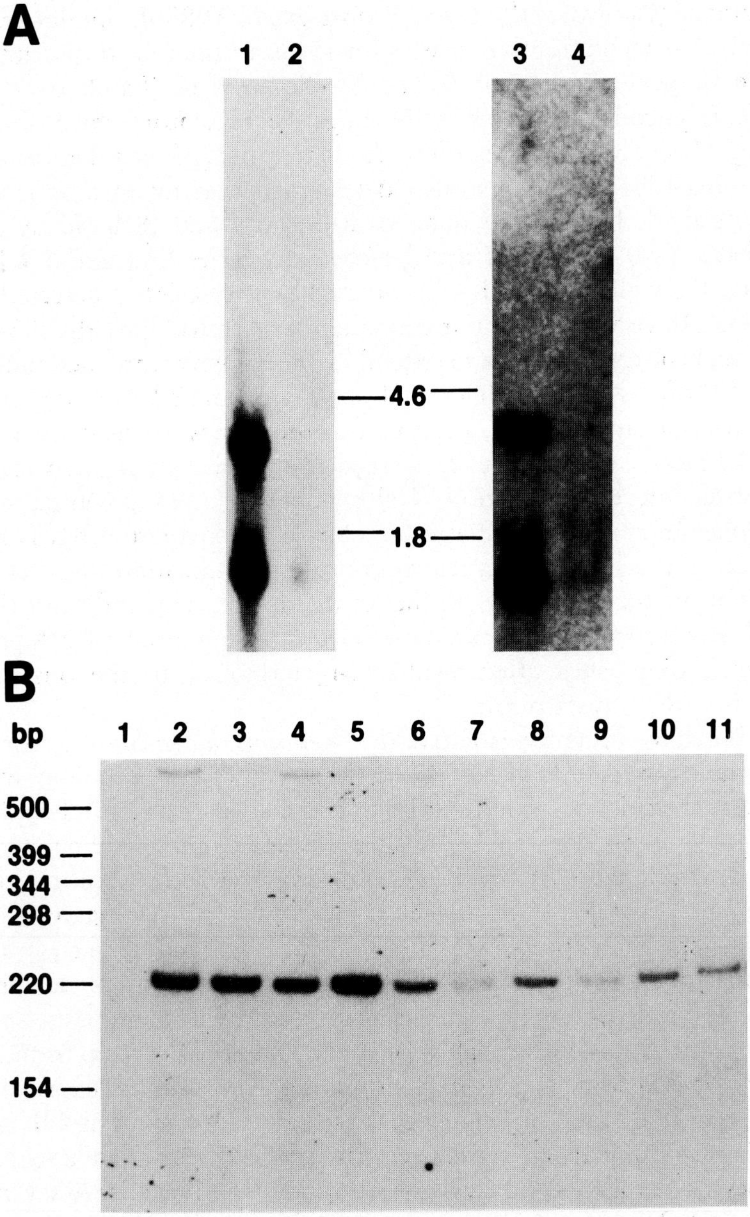
A
1
2
3
4
4.6
1.8
B
bp
1
2
3
4
5
6
7
8
9
10
11
500
399
344
298
220
154

mouse. However, in both species *mos* is part of the maternal mRNA and its disappearance precedes the onset of new embryonic transcription.

The characterization of the *mos* locus in *Xenopus* can provide an experimental *in vitro* system for determining whether it plays a role in oocyte maturation or whether it serves to imprint early embryonic transcription. The study of the c-*mos* locus in several species does reveal considerable changes in amino acid sequence and transforming activity (Schmidt *et al.*, 1988), but this may be related to its substrate or target as previously suggested (Seth *et al.*, 1987) or may indicate possible changes in function (Schmidt *et al.*, 1988). However, the similarity of the structure of the *mos* genome in different species (i.e., a single coding exon of approximately the same size with clusters of conserved regions), the expression being restricted to similar tissues in different species, and its appearance as part of the maternal mRNA in the oocytes of these species is rather strong indirect evidence for the conservation of its proto-oncogene function.

Fig. 9. Detection of chicken c-*mos* transcripts. (A) Northern analysis of RNA from ovaries and testes of chicken and quail. Twenty-five μg of total RNA was analyzed in each lane. Probe was a ^{32}P-labeled, nick-translated *Pst*I–*Pvu*II fragment from the c-*mos* coding region (Schmidt *et al.*, 1988). Ribosomal RNA is indicated. Lane 1, chicken ovary; lane 2, chicken testes; lane 3, quail ovary; lane 4, quail testes. The detection of a 1.8-kb band in quail RNA is due to background hybridization to 18S ribosomal RNA and was only observed in this particular experiment. X-ray film was exposed for 2 days. (B) S1 nuclease protection analysis. Probe was a 3′ end-labeled *Sau*3AI fragment containing 225 bp from the downstream end of the c-*mos* coding region. Lane 1, 50 μg yeast tRNA; Lane 2, 100 μg total testes RNA; Lane 3, 50 μg poly (A)+ testes RNA; Lane 4, 100 μg whole 4-day-old embryo RNA; Lane 5, 50 μg poly (A)+ 4-day-old embryo RNA; Lane 6, 100 μg 10-day-old total embryo RNA; Lane 7, 100 μg 17-day-old total embryo RNA; Lane 8, 50 μg poly (A)+ 17-day-old total embryo RNA; Lane 9, 100 μg total heart RNA; Lane 10, 100 μg total kidney RNA; Lane 11, 100 μg total spleen RNA. The X-ray film was exposed for 17 days.

References

Aaronson, S. A., and Rowe, W. P. (1970). Nonproducer clones of murine sarcoma virus-transformed BALB/3T3 cells. *Virology* **42,** 9–19.

Ben-Neriah, Y., Daley, G. Q., Mes-Masson, A. M., Witte, O. N., and Baltimore, D. (1986). The chronic myelogenous leukemia-specific P210 protein is the product of the *bcr/abl* hybrid gene. *Science* **233,** 212–214.

Blair, D. G., McClements, W. L., Oskarsson, M. K., Fischinger, P. J., and Vande Woude, G. F. (1980). Biological activity of cloned Moloney sarcoma virus DNA: Terminally redundant sequences may enhance transformation efficiency. *Proc. Natl. Acad. Sci. U.S.A.* **77,** 3504–3508.

Blair, D. G., Oskarsson, M., Wood, T. G., McClements, W. L., Fischinger, P. J., and Vande Woude, G. F. (1981). Activation of the transforming potential of a normal cell sequence: A molecular model for oncogenesis. *Science* **212,** 941–943.

Blair, D. G., Oskarsson, M. K., Seth, A., Dunn, K. J., Dean, M., Zweig, M., Tainsky, M. A., and Vande Woude, G. F. (1986). Analysis of the transforming potential of the human homolog of *mos. Cell (Cambridge, Mass.)* **46,** 785–794.

Chan, A. M.-L., King, H. W. S., Tempest, P. R., Deakin, E. A., Cooper, C. S., and Brookes, P. (1987). Primary structure of the *met* protein tyrosine kinase domain. *Oncogene* **1,** 229–233.

Cohen, J. B., Unger, T., Rechavi, G., Canaani, E., and Givol, D. (1983). Rearrangement of the oncogene c-*mos* in mouse myeloma NSI and hybridomas. *Nature (London)* **306,** 797–799.

Collins, S. J., and Groudine, M. T. (1983). Rearrangement and amplification of c-*abl* sequences in the human chronic myelogenous leukemia cell line K-562. *Proc. Natl. Acad. Sci. U.S.A.* **80,** 4813–4817.

Danos, O., and Yaniv, M. (1984). An homologous domain between the c-*mos* gene product and a papilloma virus polypeptide with a putative role in cellular transformation. *In* "Cancer Cells: Oncogenes and Viral Genes" (G. F. Vande Woude, A. J. Levine, W. C. Topp, and J. D. Watson, eds.), Vol. 2, pp. 291–294. Cold Spring Harbor Lab., Cold Spring Harbor, New York.

Di Fiore, P. P., Pierce, J. H., Kraus, M. H., Segatto, O., King, C. R., and Aaronson, S. A. (1987). *erb*B-2 is a potent oncogene when overexpressed in NIH/3T3 cells. *Science* **237,** 178–182.

Frankel, A. E., and Fischinger, P. J. (1976). Nucleotide sequence in mouse DNA and RNA specific for Moloney sarcoma virus. *Proc. Natl. Acad. Sci. U.S.A.* **73,** 3705–3709.

Gattoni, S., Kirschmeier, P., Weinstein, I. B., Escobedo, J., and Dina, D. (1982).

Cellular Moloney murine sarcoma (c-*mos*) sequences are hypermethylated and transcriptionally silent in normal and transformed rodent cells. *Mol. Cell. Biol.* **2**, 42–51.

Gattoni-Celli, S., Hsiao, W.-L. W., and Weinstein, I. B. (1983). Rearranged c-*mos* locus in a MOPC 21 murine myeloma cell line and its persistence in hybridomas. *Nature (London)* **306**, 795–796.

Goldman, D. S., Kiessling, A. A., Millette, C. F., and Cooper, G. M. (1987). Expression of c-*mos* RNA in germ cells of male and female mice. *Proc. Natl. Acad. Sci. U.S.A.* **84**, 4509–4513.

Gonzatti-Haces, M., Seth, A., Park, M., Copeland, T., Oroszlan, S., and Vande Woude, G. F. (1988). Characterization of the *tpr–met* oncogene p65 and the *met* proto-oncogene cell surface p140 tyrosine kinases. *Proc. Natl. Acad. Sci. U.S.A.* **85**, 21–25.

Goyette, M., Petropoulos, C. J., Shank, P. R., and Fausto, N. (1984). Regulated transcription of c-Kr-*ras* and c-*myc* during compensatory growth of rat liver. *Mol. Cell. Biol.* **4**, 1493–1498.

Hayman, M. J., Ramsay, G. M., and Savin, K. (1983). Identification and characterization of the avian erythroblastosis virus *erbB* gene product as a membrane glycoprotein. *Cell (Cambridge, Mass.)* **32**, 579–588.

Heisterkamp, N., Stephenson, J. R., Groffen, J., Hansen, P. I., de Klein, A., Bartram, C. R., and Grosfeld, G. (1983). Localization of the c-*abl* oncogene adjacent to a translocation break point in chronic myelocytic leukaemia. *Nature (London)* **306**, 239–242.

Hudziak, R. M., Schlessinger, J., and Ullrich, A. (1987). Increased expression of the putative growth factor receptor $p185^{HER2}$ causes transformation and tumorigenesis of NIH 3T3 cells. *Proc. Natl. Acad. Sci. U.S.A.* **84**, 7159–7163.

Jones, M., Bosselman, R. A., van der Hoorn, F. A., Berns, A., Fan, H., and Verma, I. (1980). Identification and molecular cloning of Moloney mouse sarcoma virus-specific sequences from uninfected mouse cells. *Proc. Natl. Acad. Sci. U.S.A.* **77**, 2651–2655.

Keshet, E., Rosenberg, M. P., Mercer, J. A., Propst, F., Vande Woude, G. F., Jenkins, N. A., and Copeland, N. G. (1988). Developmental regulation of ovarian-specific *mos* expression. *Oncogene* **2**, 235–240.

Konopka, J. B., Watanabe, S. M., and Witte, O. N. (1984). An alteration of the human c-*abl* protein in K562 leukemia cells unmasks associated tyrosine kinase activity. *Cell (Cambridge, Mass.)* **37**, 1035–1042.

Lambert, P. F., Spalholz, B. A., and Howley, P. M. (1987). A transcription repressor encoded by BPV-1 shares a common carboxy-terminal domain with the E2 transactivator. *Cell (Cambridge, Mass.)* **50**, 69–78.

Maguire, R. T., Robins, T. S., Thorgeirsson, S. S., and Heilman, C. A. (1983).

Expression of cellular *myc* and *mos* genes in undifferentiated B cell lymphomas of Burkitt and non-Burkitt types. *Proc. Natl. Acad. Sci. U.S.A.* **80,** 1947–1950.

Martin-Zanca, D., Hughes, S. H., and Barbacid, M. (1986). A human oncogene formed by the fusion of truncated tropomyosin and protein tyrosine kinase sequences. *Nature (London)* **319,** 743–748.

Maxwell, S. A., and Arlinghaus, R. B. (1985). Serine kinase activity associated with Moloney murine sarcoma virus-124-encoded $p37^{mos}$. *Virology* **143,** 321–333.

Moloney, J. B. (1966). A virus-induced rhabdomyosarcoma in mice. *Natl. Cancer Inst. Monogr.* **22,** 139–142.

Muller, R., Slamon, D. J., Tremblay, J. M., Cline, M. J., and Verma, I. M. (1982). Differential expression of cellular oncogenes during pre- and postnatal development. *Nature (London)* **299,** 640–644.

Mutter, G. L., and Wolgemuth, D. J. (1987). Distinct developmental patterns of c-*mos* proto-oncogene expression in female and male mouse germ cells. *Proc. Natl. Acad. Sci. U.S.A.* **84,** 5301–5305.

Oskarsson, M., McClements, W. L., Blair, D. G., Maizel, J. V., and Vande Woude, G. F. (1980). Properties of a normal mouse cell DNA sequence (sarc) homologous to the *src* sequence of Moloney sarcoma virus. *Science* **207,** 1222–1224.

Papkoff, J., Verma, I. M., and Hunter, T. (1982). Deletion of a transforming gene product in cells transformed by Moloney murine sarcoma virus. *Cell (Cambridge, Mass.)* **29,** 417–426.

Papkoff, J., Nigg, E. A., and Hunter, T. (1983). The transforming protein of Moloney murine sarcoma virus is a soluble cytoplasmic protein. *Cell (Cambridge, Mass.)* **33,** 161–172.

Park, M., Dean, M., Cooper, C. S., Schmidt, M., O'Brien, S. J., Blair, D. G., and Vande Woude, G. F. (1986a). Mechanism of *met* oncogene activation. *Cell (Cambridge, Mass.)* **45,** 895–904.

Park, M., Gonzatti-Haces, M., Dean, M., Blair, D. G., Testa, J. R., Bennett, D. D., Copeland, T., Oroszlan, S., and Vande Woude, G. F. (1986b). The *met* oncogene: A new member of the tyrosine kinase family and a marker for cystic fibrosis. *Cold Spring Harbor Quant. Symp. Biol.* **51,** 967–975.

Park, M., Dean, M., Kaul, K., Braun, M. J., Gonda, M. A., and Vande Woude, G. F. (1987). Sequence of *met* proto-oncogene cDNA has features characteristic of the tyrosine kinase family of growth factor receptors. *Proc. Natl. Acad. Sci. U.S.A.* **84,** 6379–6383.

Paules, R. S., Propst, F., Dunn, K. J., Blair, D. G., Kaul, K., Palmer, A. E., and Vande Woude, G. F. (1988). Primate c-*mos* proto-oncogene structure and expression: Transcription initiation both upstream and within the gene in a tissue-specific manner. *Oncogene* **3,** 59–68.

Privalsky, M. L., and Bishop, J. M. (1984). Subcellular localization of the v-*erb-B* protein, the product of a transforming gene of avian erythroblastosis virus. *Virology* **135**, 356–368.

Propst, F., and Vande Woude, G. F. (1985). c-*mos* proto-oncogene transcripts are expressed in mouse tissues. *Nature (London)* **315**, 516–518.

Propst, F., Rosenberg, M. P., Iyer, A., Kaul, K., and Vande Woude, G. F. (1987). c-*mos* proto-oncogene RNA transcripts in mouse tissues: Structural features, developmental regulation and localization in specific cell types. *Mol. Cell. Biol.* **7**, 1629–1637.

Propst, F., Rosenberg, M. P., Oskarsson, M. K., Russell, L. B., Nguyen-Huu, M. C., Nadeau, J., Jenkins, N. A., Copeland, N. G., and Vande Woude, G. F. (1988a). Genetic analysis and developmental regulation of testis-specific RNA expression of *Mos*, *Abl*, actin and *Hox-1.4*. *Oncogene* **2**, 227–233.

Propst, F., Rosenberg, M. P., and Vande Woude, G. F. (1988b). Proto-oncogene expression in germ cell development. *Trends Genet.* **4**, 183–187.

Rechavi, G., Givol, D., and Canaani, E. (1982). Activation of a cellular oncogene by DNA rearrangement: Possible involvement of an IS-like element. *Nature (London)* **300**, 607–611.

Rhim, J. S., Park, D. K., Arnstein, P., Huebner, R. J., and Weisburger, E. K. (1975). Transformation of human cells in culture by *N*′-methyl-*N*′-nitro-*N*-nitrosoguanidine. *Nature (London)* **256**, 751–753.

Sagata, N., Oskarsson, M., Copeland, T., Brumbaugh, J., and Vande Woude, G. F. (1988). The c-*mos* proto-oncogene product functions during meiotic maturation in *Xenopus* oocytes. *Nature (London)* **335**, 519–525.

Schmidt, M., Oskarsson, M. K., Dunn, J. K., Blair, D. G., Hughes, S., Propst, F., and Vande Woude, G. F. (1988). Characterization of the chicken homolog of the *mos* proto-oncogene. *Mol. Cell. Biol.* **8**, 923–929.

Seth, A., and Vande Woude, G. F. (1985). Nucleotide sequence and biochemical activities of the Moloney murine sarcoma virus strain HT-1 *mos* gene. *J. Virol.* **56**, 144–152.

Seth, A., Priel, E., and Vande Woude, G. F. (1987). Nucleoside triphosphate-dependent nucleic-acid-binding properties of *mos* protein. *Proc. Natl. Acad. Sci. U.S.A.* **84**, 3560–3564.

Sherr, C. J., Rettenmier, C. W., Sacca, R., Roussel, M. F., Look, A. T., and Stanley, E. R. (1985). The c-*fms* proto-oncogene product is related to the receptor for the mononuclear phagocyte growth factor, CSF-1. *Cell (Cambridge, Mass.)* **41**, 665–676.

Ullrich, A., Coussens, L., Hayflick, J. S., Dull, T. J., Gray, A., Tam, A. W., Lee, J., Yarden, Y., Libermann, T. A., Schlessinger, J., Downward, J., Mayes, E. L. V., Whittle, N., Waterfield, M. D., Seeburg, P. H. (1984). Human epidermal growth factor receptor cDNA sequence and aberrant expression of the amplified gene in A431 epidermoid carcinoma cells. *Nature (London)* **309,** 418–425.

Van Beveren, C., van Straaten, F., Galleshaw, J. A., and Verma, I. M. (1981). Nucleotide sequence of the genome of a murine sarcoma virus. *Cell (Cambridge, Mass.)* **27,** 97–108.

van der Hoorn, F. A., and Firzlaff, J. (1984). Complete c-*mos* (rat) nucleotide sequence: Presence of conserved domains in c-*mos* proteins. *Nucleic Acids Res.* **12,** 2147–2156.

Vande Woude, G. F., Oskarsson, M., McClements, W. L., Enquist, L. W., Blair, D. G., Fischinger, P. J., Maizel, J. V., and Sullivan, M. (1980). Characterization of integrated Moloney sarcoma proviruses and flanking host sequences cloned in bacteriophage λ. *Cold Spring Harbor Symp. Quant. Biol.* **44,** 735–745.

Watson, R., Oskarsson, M., and Vande Woude, G. F. (1982). Human DNA sequence homologous to the transforming gene (*mos*) of Moloney murine sarcoma virus. *Proc. Natl. Acad. Sci. U.S.A.* **79,** 4078–4082.

Wood, T. G., Blair, D. G., and Vande Woude, G. F. (1983a). Moloney sarcoma virus: Analysis of RNA and DNA structure in cells transformed by subgenomic proviral DNA fragments. *In* "Perspectives on Genes and the Molecular Biology of Cancer" (D. L. Robberson and G. F. Saunders, eds.), pp. 299–306. Raven Press, New York.

Wood, T. G., McGeady, M. L., Blair, D. G., and Vande Woude, G. F. (1983b). Long terminal repeat enhancement of v-*mos* transforming activity: Identification of essential regions. *J. Virol.* **46,** 726–736.

11

Replication and Pathogenesis of the Human Retrovirus Relevant to Drug Design

WILLIAM A. HASELTINE, JOSEPH SODROSKI, AND
ERNEST TERWILLIGER

Department of Pathology
Harvard Medical School
Boston, Massachusetts

THE REGULATION OF PROLIFERATION AND
DIFFERENTIATION IN NORMAL AND NEOPLASTIC CELLS

I. Introduction

Infection with the human immunodeficiency virus (HIV) initiates a progressive degenerative disease of the immune and central nervous systems. What follows is a summary of the experiments from this laboratory designed to develop an understanding of the disease in terms of the life cycle and biology of the virus. The results can be summarized by the following statement:

AIDS is caused by a retrovirus that establishes controlled, persistent infection in patients that is selectively cytotoxic to the helper subset of T cells and that is designed to evade the immune response.

II. Controlled Infection

Infection by HIV is often followed by a long asymptomatic period (Goedert *et al.*, 1984). During the asymptomatic state, a low but significant level of virus is found in the blood (Eyster *et al.*, 1985; Goedert *et al.*, 1986). Later on in the course of the disease the level of virus usually rises. Such controlled replication may persist for many years (Fig. 1). By contrast, growth of HIV in cultures of CD4+ cells is very rapid, killing all the cells in a week (Klatzmann *et al.*, 1984; Popovic *et al.*, 1984; Sodroski *et al.*, 1986a; Barre-Sinoussi *et al.*, 1983).

Clearly, control of virus replication is central to the disease process. One aspect of the control of HIV replication may be the immune response to the virus. However, studies of the virus show that the virus itself has elaborate genetic control mechanisms that govern why, where, and how much virus is made. What emerges

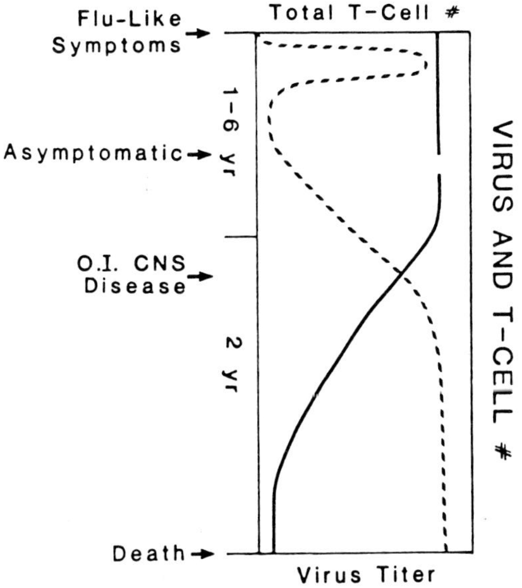

Fig. 1. Idealized description of changes in viral titers and levels of circulating lymphocytes in HIV-infected individuals over time. ---, Virus, ———, lymphocytes.

from these molecular studies is the picture of a virus that contains counterbalanced regulatory pathways; a virus that is able either to replicate exuberantly or to lie dormant, not to replicate at all, until an appropriate signal is received. We are beginning to understand the different virus replication states in terms of the structure and function of the genes and regulatory elements of the AIDS virus.

The AIDS virus differs from other retroviruses previously investigated in the presence of an elaborate superstructure of regulatory components that govern virus replication. The simplest retroviruses contain genes that encode the virion core component (*gag*), replication functions (*pol*), and the envelope glycoprotein (*env*). These simple retroviruses replicate constitutively. The only regulation of the growth of these viruses is that observed by the replication state of the host cell itself. Nonreplicating cells cannot be productively infected as a consequence of failure of the provirus

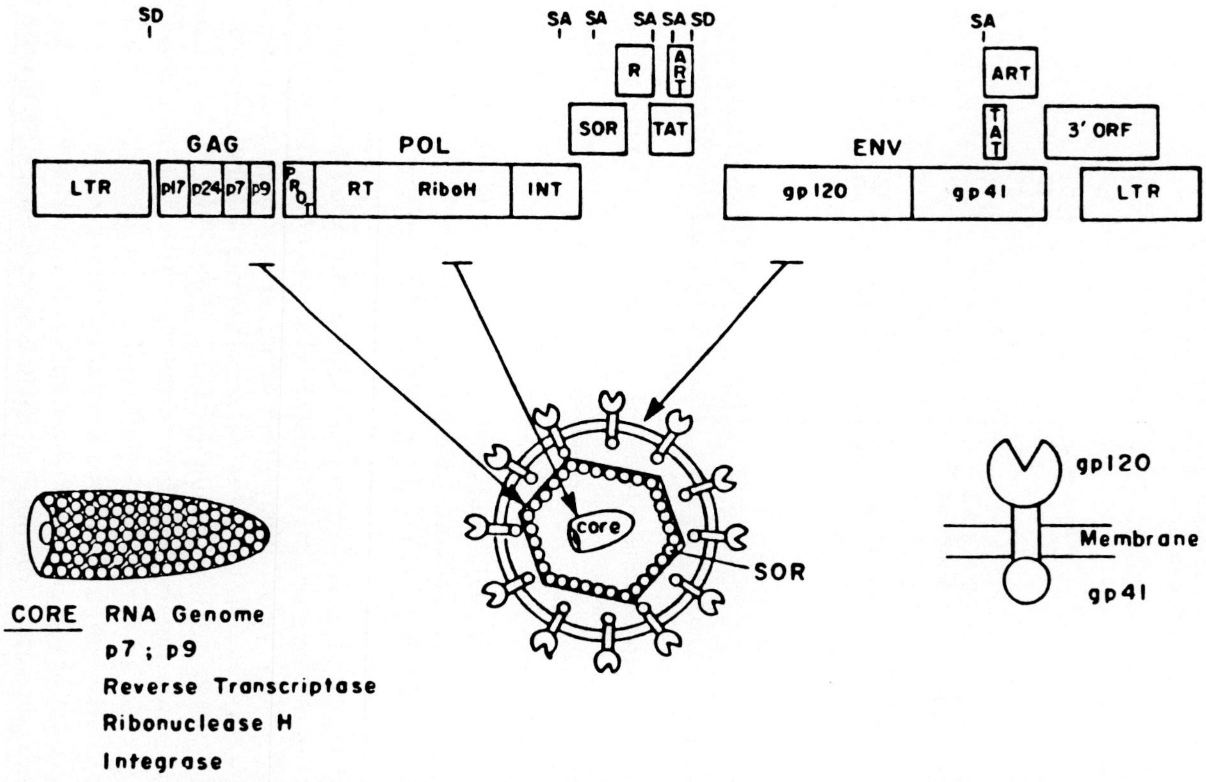

Fig. 2. Genome map of HIV, and schematic diagram of the incorporation of the different viral products into particles. SD, splice donor; SA, splice acceptor.

to integrate. Resting cells that contain integrated provirus do not produce virus as a consequence of failure of the resting cells to initiate RNA transcription from the proviral DNA. Aside from such cellular control of virus infection, no additional virally encoded regulation of replication has been described for these simple retroviruses.

HIV, by contrast, encodes at least four proteins and five cis-acting regulatory sequences that govern virus replication. Some of these regulatory elements allow viral growth whereas others accelerate growth (Fig. 2).

A. The Long Terminal Repeat

Several of these regulatory elements are located on the viral long terminal repeat (LTR), that segment of the virus flanking the proviral genes (Fig. 3). As for all retroviruses, the LTR serves a critical role in the control of virus gene expression and replication. The LTR contains sequences that specify RNA initiation, the first process in the expression of virus information whereby viral DNA is transcribed into the corresponding RNA. The initial RNA product extends the full length of the genome. It is later spliced in the nucleus to produce smaller RNAs from which some of the viral proteins are made.

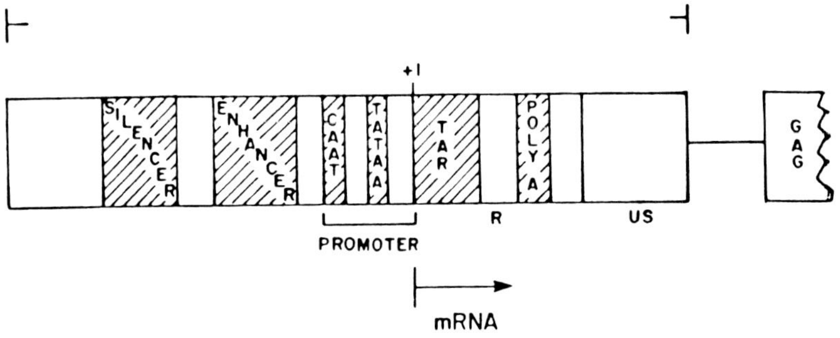

Fig. 3. Schematic diagram of the long terminal repeat of HIV. Transcription begins at +1.

The LTR contains sequences near the site of initiation that locate the precise site of RNA initiation as well as sequences located 5′ to the start site, called *enhancer sequences*, that facilitate initiation (Fig. 3) (Rosen *et al.*, 1985; Starcich *et al.*, 1985; Muesing *et al.*, 1987; Franza *et al.*, 1987).

There is no apparent cell preference for the function of the HIV LTR. Once in a cell, the LTR works regardless of the tissue of origin and even the species of origin (Sodroski *et al.*, 1985a; Peterlin *et al.*, 1986).

It is notable that the LTR of HIV works poorly as a promoter regardless of cell type in uninfected cells. Compared to most other viral promoters, the HIV promoter is very weak and directs only a low level of viral mRNA synthesis in uninfected cells. The low level of RNA synthesis may be one of the mechanisms by which HIV controls the rate of infection.

One regulatory element not found in the LTR of most other retroviruses is the cis-acting negative regulator of transcription called the *negative regulatory element*, or NRE. This sequence functions as a "silencer" (Brand *et al.*, 1985; Rosen *et al.*, 1985).

The low level of initial RNA transcription in HIV-infected cells may be attributed, at least in part, to the balance between the positive contribution of the enhancer and the negative contribution of the silencer.

B. The 3′ *orf* Gene

HIV encodes a protein, 3′ *orf*, that negatively regulates the rate of replication. The product of the 3′ *orf* gene is a 27-kDa protein that is myristilated and located predominantly in the cytoplasm (Allan *et al.*, 1985b). Purified bacterially expressed 3′ *orf* protein has also been reported to possess GTPase and autophosphorylation activities (Guy *et al.*, 1987). Viruses lacking the 3′ *orf* gene replicate more rapidly in CD4+ cells than do viruses that express this protein (Terwilliger *et al.*, 1986; Fisher *et al.*, 1986; Luciw *et al.*, 1987). In effect, the 3′ *orf* protein serves as a "brake" to virus replication and contributes to the controlled replication of the virus. The 3′ *orf* gene is not required for virus replication as viruses incapable of producing the 3′ *orf* gene product replicate and kill CD4+ cells.

Some isolates of HIV are defective in 3′ *orf* expression. It is likely that these isolates represent *in vitro* selection of strains that replicate rapidly and that such strains do *not* represent natural virus isolation, as patients often make antibodies to the 3′ *orf* protein. The 3′ *orf* gene is present in all HIV-1 strains as well as HIV-2 isolates.

The 3′ *orf* gene may slow down virus gene expression during the initial stage of infection to permit establishment of latent, nonproductive infection. It has been reported that expression of 3′ *orf* protein in a CD4+ cell line down-regulates expression of CD4 antigen (Guy *et al.*, 1987). The action of the 3′ *orf* gene may help to explain the low levels of free virus seen in most infected people.

C. The *sor* Gene

HIV encodes a protein, *sor,* that accelerates viral growth. The *sor* product is a 23-kDa protein that is located predominantly in the cytoplasm of infected cells. The mutants defective for *sor* produce virions that replicate more slowly than does the wild-type virus. *Sor*-defective virions can establish productive cytopathic infections in CD4+ cells but the virus replicates more slowly than does the wild-type virus (Sodroski *et al.*, 1986b). The *sor* gene is not required for replication, infectivity, or cytopathic effect of HIV but is required for rapid growth. Recent studies suggest that the infectivity of the virions of *sor*-defective virus for CD4+ cells is reduced as compared to wild-type virus (Fisher *et al.*, 1987; Strebel *et al.*, 1987). The extent of the defect depends on the recipient cell line. However, cell-free *sor*-defective virus can infect some CD4+ cells in culture (T. Dorfman, J. Sodroski, and W. Haseltine, unpublished observations).

D. The Trans-Activator (*tat*) Gene

The *tat* gene of HIV encodes the *trans-activator* function of the virus (Sodroski *et al.*, 1985b; Arya *et al.*, 1985). The product of this gene is a 14-kDa protein (Goh *et al.*, 1986; Wright *et al.*, 1986) located primarily in the nucleus of infected cells (Hauber *et al.*, 1987). The *tat* protein permits very high levels of viral protein to be made despite the low level of transcription directed by the HIV LTR in uninfected cells (Rosen *et al.*, 1985; Sodroski *et al.*, 1985a). The *tat* gene is essential for virus replication (Dayton *et al.*, 1986; Fisher *et al.*, 1986b). The product of the *tat* gene acts on the viral LTR to increase the rate of production of *tat* protein and of other viral proteins. It acts as a positive feedback regulator. The more *tat* protein made, the more rapidly the *tat* protein itself is made. The *tat* product permits explosive growth of HIV under appropriate conditions.

The critical response element needed for *tat* activity is located from +1 to +80 within the viral LTR and is called TAR for *trans-acting responsive region* (Rosen *et al.*, 1985, 1986a; Sodroski *et al.*, 1985b; Wright *et al.*, 1986). The mode of action of the *tat* protein is not fully understood. On balance, the strongest evidence indicates that the *tat* protein greatly increases the efficiency of translation of mRNA species in the absence of the *tat* product (Rosen *et al.*, 1986a; Feinberg *et al.*, 1986; Wright *et al.*, 1986). However, it has been noted that steady-state levels of RNA that contain 5′ TAR sequences increase in the presence of the *tat* protein (Cullen, 1986; Wright *et al.*, 1986; Peterlin *et al.*, 1986). It is also reported that rates of HIV LTR-initiated RNA synthesis increased in the presence of the *tat* protein, as judged by run-on transcription (Hauber *et al.*, 1987). The results of both of the experiments described above could be explained by instability in the absence of the *tat* protein of mRNAs that initiate within the HIV LTR TAR sequences. Indeed, a recent measurement of the level of synthesis of the 5′-most RNA sequences of TAR-initiated RNA revealed no differences in the rate of synthesis of the 5′ RNA in the presence or absence of the *tat* product, as judged by run-on synthesis. In these experiments, the level of RNA distal to the TAR sequences was greater in the presence of the *tat* protein as compared to the level of such synthesis in the absence of the *tat* product. Kao *et al.* (1987) attribute such differences to a transcription antitermination activity of the *tat* product. This interpretation of the result is flawed by the observation that the expression of heterologous gene products produced by the HIV LTR does not increase upon deletion of the proposed transcription attenuation sequences (Rosen *et al.,* 1985). However, the experiment of Kao *et al.* indicates that differences in transcription initiation do not play an important role in *tat* action. Overall, these experiments support the notion that the *tat* gene product acts primarily to regulate the expression of viral protein expression posttranscriptionally.

The *tat* protein is located primarily in the nucleus, and there is some evidence for preferential concentration within the nucleolus (Hauber *et al.* 1987). We speculate that *tat* protein facilitates transport of TAR-containing RNAs from the nucleus to the cytoplasm.

Noteworthy is the observation that 5′ sequences of HIV mRNA inhibit the translation of such RNA *in vivo* in *Xenopus* oocytes and in cells transfected with plasmids that yield mRNA species that initiate with the HIV 5′ mRNA sequences.

The combined effect of the *tat*–TAR positive regulatory loop is to permit explosive growth of the virus when the conditions are right. The lethal effects of the virus as well as the transmissibility of the virus from person to person are likely to depend upon the unusual ability of the virus to undergo *tat*-mediated explosive growth.

E. The *art* Gene

HIV also encodes the *antirepression transactivator* (*art*) protein that serves as a genetic switch permitting differential synthesis of regulatory and structural genes (Fig. 4). The *art* protein is a 20-kDa protein located predominantly in the nucleus of infected

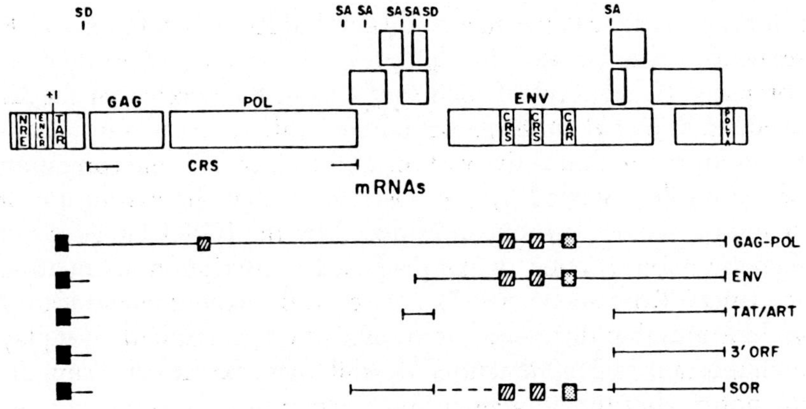

Fig. 4. Locations of the cis-acting elements involved in regulation by the *art* gene product within the different messenger RNA species of HIV. Dashed line for the *sor* mRNA reflects current uncertainty concerning its precise structure. CRS, cis-acting repression sequence; CAR, cis-acting antirepression sequence.

cells. The *art* gene is essential for virus replication (Sodroski *et al.*, 1986a; Feinberg *et al.*, 1986; Knight *et al.*, 1987; Terwilliger *et al.*, 1988). Without the *art* gene function only regulatory proteins, not the proteins that make up the virus particle, are made. There are three components to the *art* regulatory pathway: the *art* protein itself, sequences called *repression sequences*, and CAR (cis-acting antirepression) sequences (Rosen *et al.*, 1988). The *art* protein is located predominantly in the nucleus (C. A. Rosen, unpublished observation). We propose the following hypothesis for *art* activity. The CRS sequences signal retention of the RNAs which contain such sequences in the nucleus. Retention of the RNA in the nucleus prevents access of the mRNAs to the ribosomes and therefore prevents expression of proteins encoded by such mRNAs. However, upon removal of cis-acting repression sequences (CRS) from the RNA by splicing, the processed RNA devoid of the CRS sequences may exit. This is the mechanism proposed for expression of regulatory proteins in the absence of the *art* protein. The action of the *art* protein on CAR is proposed to specify rapid exit from the nucleus to the cytoplasm of mRNAs where such RNA can be translated or incorporated into virion particles. It is proposed that the *art* activity on CAR overrides the nuclear retention signal of the CRS elements.

This hypothesis would account for the effect of the *art* protein on both the translation of viral proteins and on the ratio of spliced to unspliced viral mRNAs, which increases in the absence of a functional *art* protein. In the absence of the art function, the tendency would be for full-length RNA transcripts to be retained in the nucleus until the CRS sequences are removed by splicing. The *art* activity and CAR would permit expression of messages which contain CRS sequences (i.e., the *gag, pol,* and *env* messages).

Viewed from this perspective, both *tat* and *art* can be considered to specify mRNA nuclear-cytoplasmic transport functions.

The *art* protein is likely to play a key role in the early steps of infection that determine lytic versus latent growth. If *art* is active, then the virus structural proteins are made and cells may be killed. If *art* is not active, then a latent state can be established because virus structural proteins that may be lethal to cells are not made.

The same *art* switch probably plays a key role in the timing of lytic infections, as well as release from the latent state, to permit large bursts of virus protein. The virus proteins, particularly the *env* gene product, are lethal to CD4+ cells. If the *env* gene product is made early, CD4+ cells may die before much virus is made. In activation from the latent state, the *art* activity would permit accumulation of some protein and some viral RNAs, so that when *art* is eventually activated, virion production is explosive.

Undoubtedly these counterbalanced genetic elements of HIV—the opposed silencer–enhancer combination in the LTR, the opposed activities of *sor* and 3′ *orf*, and the "pas de deux" of the *tat*–TAR and *art*–CAR–CRS regulatory loops—play a central role in the pathogenesis of AIDS as well as the transmission of the virus.

III. Selective Cytotoxicity

HIV infection is selectively cytotoxic. Infection by HIV kills actively replicating CD4+ helper T cells (Barre-Sinoussi *et al.*, 1983; Klatzmann *et al.*, 1984; Popovic *et al.*, 1984). This cytopathic effect is probably the primary cause for the immune deficiency in AIDS patients. However, HIV infects other types of cells such as monocytes (Gartner *et al.*, 1986), macrophages (Salahuddin *et al.*, 1986), and glial cells (Chiodo *et al.*, 1987; Cheng-Meyer *et al.*, 1987), with little or no lethal effect. Persistent infection of such cell types without cell death probably accounts for the low level of virus found in infected patients and may also contribute to the continued ability of such patients to transmit the infection during the asymptomatic phase of the disease.

We have offered a simple hypothesis to explain the selective cytotoxic activity of HIV. The hypothesis is summarized by the equation

$$[\text{gp120–gp41}] \times [\text{CD4}] \times [\text{Fusion Factor}] = \text{Cell death}$$

This hypothesis has now been confirmed by several researchers and is now generally accepted. The predictions of this model have been confirmed by direct test.

Simply put, the hypothesis holds that the death of an infected cell depends upon the simultaneous high-level expression of the envelope glycoprotein and CD4 molecule that serves as a receptor for viral infections (McDougal *et al.*, 1986), as well as the presence of a cellular factor, or *fusion factor,* required for envelope-mediated fusion. Cell death occurs in CD4+ cells that are actively replicating and producing large amounts of the virus. The concentration of CD4 is high on CD4+ T helper cells (20,000–60,000 copies per cell) and the concentration of the *env* protein is also high. Noncytopathic productive infection occurs in monocytes and macrophages, in which the concentration of CD4 is low, less than 2000 per cell. This number of receptors is large enough to permit infection by the virus but too low to permit envelope-mediated cell killing. The requirement for a fusion factor is inferred from experiments in which some cell lines into which the CD4 gene has been transferred have been shown to permit binding of HIV envelope but not virus entry.

The interaction between the envelope glycoprotein and the CD4 molecule is important not only for cell killing but also for virus infection. The key components of the reaction that permit viral entry and also determine virus killing are diagrammed (Fig. 5). The *env* protein is composed of two subunits. Gp120, the exterior glycoprotein, is located outside the cell membrane, and gp41 is anchored to the membrane with portions extending both outside and inside the membrane of the infected cell or virus particle. The

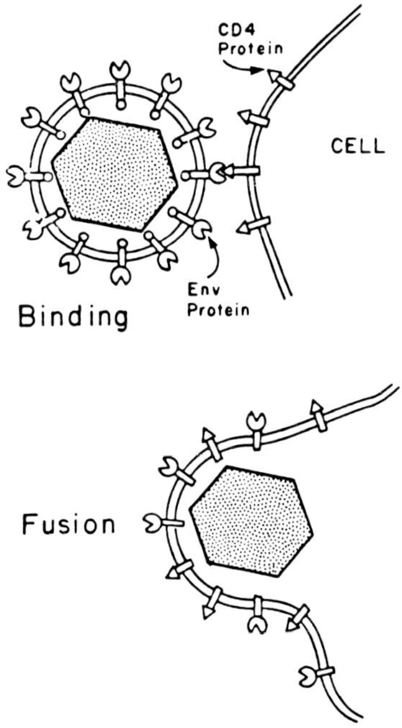

Fig. 5. Diagrammatic representation of the process of virus entry into CD4+ cells. Tight binding reaction between the exterior glycoprotein gp120 and CD4 brings opposing membranes into close proximity, followed by a fusion reaction, mediated by the transmembrane protein gp41, which merges the two membranes together.

CD4 molecule is bound to the surface of the uninfected cell membrane. A very tight binding reaction occurs between the exterior glycoprotein, gp120, and CD4. This tight binding draws the two membranes together. The envelope protein then initiates a fusion reaction. It is the CD4 binding and subsequent fusion reaction that permits virus to enter the cell. This reaction also enables infected cells to fuse with surrounding uninfected CD4+ cells (Fig. 6). It is probably this reaction that also destroys infected cells that express high concentrations of both CD4 and envelope protein.

The binding and fusion reaction are of such importance to the infection and lethal properties of HIV that we have investigated this reaction in some detail (Kowalski *et al.*, 1987). Our approach was to introduce mutations into the envelope glycoprotein to observe the effects on binding and fusion reactions. The mutants could be divided into several groups that define the functions

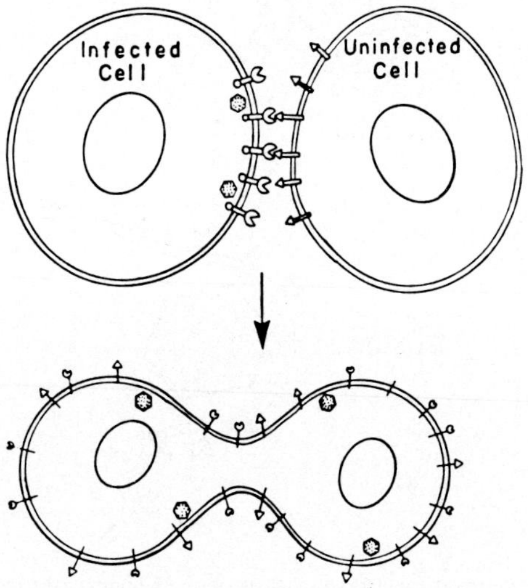

Fig. 6. Diagrammatic representation of the process of syncytia formation. Same process of gp120–CD4 binding followed by membrane fusion as illustrated in Fig. 5, this time between gp120 in membrane of infected cell and CD4 molecules of uninfected cells.

necessary for the binding and fusion reactions. Some mutants affected binding of gp120 to CD4. Some mutants inhibited the fusion reaction but permitted gp120 binding of CD4. Some mutants inhibited anchorage of gp41 to the membrane. Some mutants destroyed gp120–gp41 association. Some mutants prevented cleavage of the gp120–gp41. Fortunately, the mutants clustered into discrete areas of the protein and permitted definition of the functional regions in terms of specific amino acid sequences.

A. CD4 Binding Regions

Mutants that affect CD4 binding are located in three separate regions near the carboxy terminus of gp120 (Fig. 7). All of the regions are highly conserved among HIV-1 strains and between HIV-1 and HIV-2. The regions are separated by hypervariable regions, and changes in these regions do not affect gp120 binding to CD4. Additional experiments also indicate that these regions bind CD4 (Laskey *et al.*, 1987). Antisera raised to a conserved peptide to one of these regions inhibits gp120 binding of CD4.

We propose that these three noncontiguous sequences are brought together by the tertiary structure of the protein to form a CD4 binding pocket that is interdigitated with hypervariable regions that form the folds between the faces of the binding pocket.

B. Fusion

Mutants in the amino terminus of gp41 permit normal synthesis and processing of the *env* protein and permit gp120 binding to CD4 but inhibit the fusion reaction (Fig. 8). The hydrophobic region affected is analogous to those of other enveloped viruses involved in fusion reactions. We propose that the hydrophobic amino terminus of the gp41 is inserted into the opposing membrane and disrupts the lipid bilayer, thus initiating a fusion event.

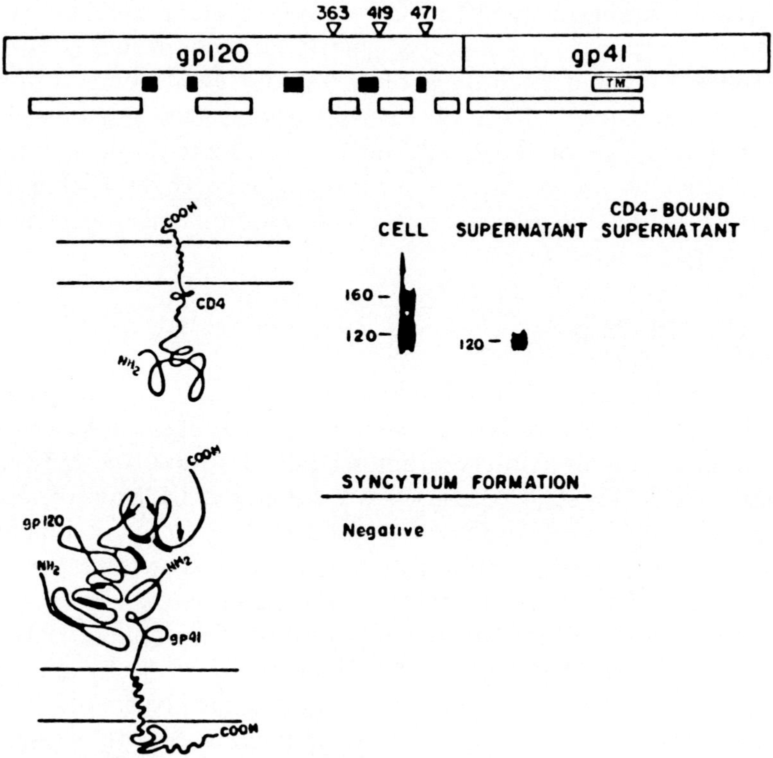

Fig. 7. Localization of HIV envelope mutations disrupting binding of gp120 to CD4. Changes within any of three short conserved domains in the coding sequences for the carboxy terminus of gp120 eliminate the ability of the protein to bind the CD4 molecule. Triangles designate points at which mutations were made, numbers above triangles designate amino acid residue numbers. White bars below diagram indicate highly conserved regions of the envelope protein. Solid bars indicate hypervariable regions. Three autoradiography bars illustrate the phenotype observed for this class of mutations. CD4+ lymphocytes transfected with expressors for the mutant envelope proteins were metabolically labeled. Immunoprecipitations of labeled proteins with AIDS patient antiserum were carried out either on the transfected cells themselves (cell), the medium in which the cells were labeled (supernatant), or fresh CD4+ lymphocytes after incubation with the labeled supernatant (CD4-bound supernatant).

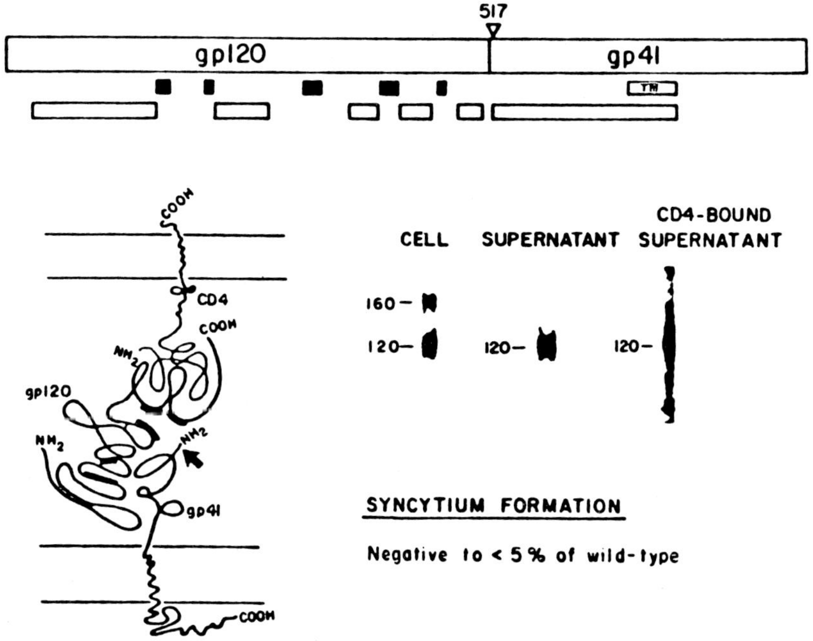

Fig. 8. Localization of HIV envelope mutations affecting the membrane fusion function. Mutations within a single hydrophobic domain in the amino terminus of gp41 produce *env* proteins with normal CD4 binding activity, but little or no syncytia-forming activity. For abbreviations, see Fig. 7.

C. gp120–gp41 Cleavage

Some mutants prevent gp160 cleavage (Fig. 9). Such mutants do not bind or fuse to CD4+ cells. We have also shown that wild-type, unmutated gp160 does not bind to CD4. We propose that the cleavage reaction is necessary to release the carboxy terminus of gp120 and the amino terminus of gp41 so that they may assume the configuration required for the binding and fusion reactions, respectively.

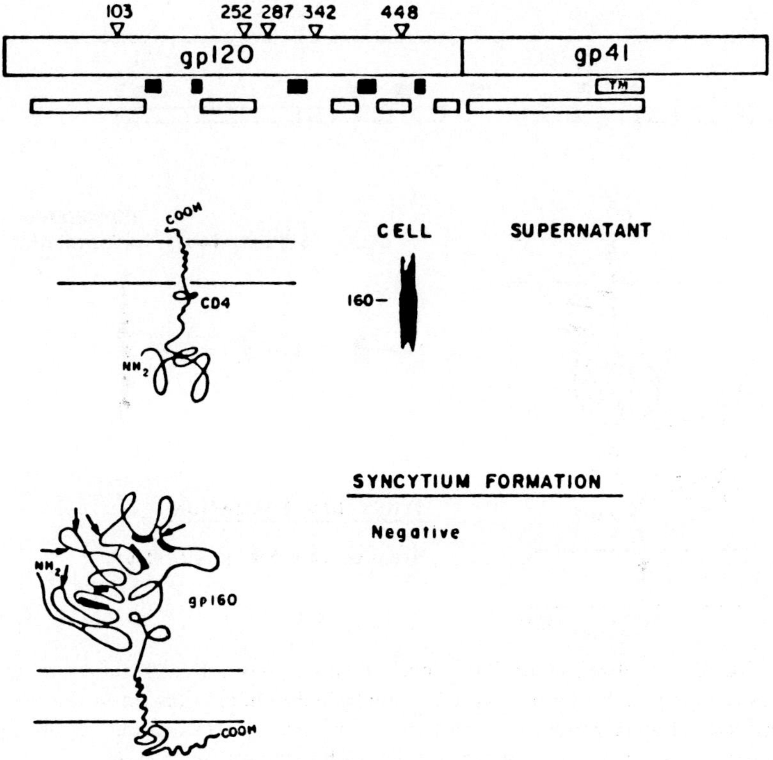

Fig. 9. Widely scattered mutations within the HIV envelope gene all result in defective processing of the *env* product so that no cleavage of the gp160 to form gp120 and gp41 occurs. For abbreviations, see Fig. 7.

D. gp120–gp41 Association

Some mutants weaken the association of gp120 and gp41. These mutants cluster at the amino terminus of both proteins (Fig. 10). Such mutants produce normal levels of proteins but gp120 is not attached to the cell surface and is released into the culture medium. Such mutants do not initiate fusion events. We propose that the gp120 and gp41 proteins associate via a series of noncovalent interactions between the amino terminals of both

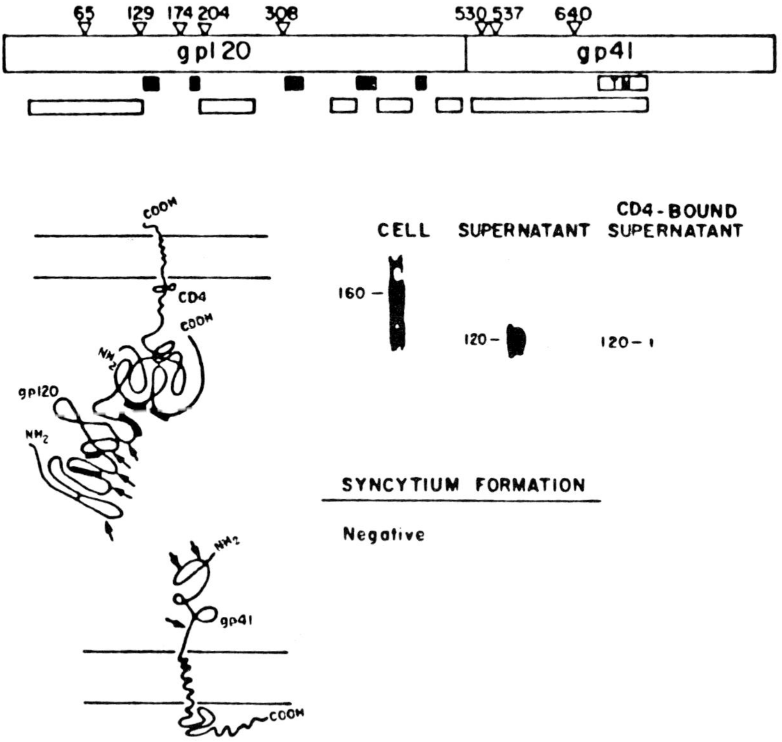

Fig. 10. Mutations within the amino termini of either the HIV envelope gp120 or gp41 result in loss of the normal noncovalent association between the two proteins. For abbreviations, see Fig. 7.

proteins and that the gp120 association with gp41 is critical. Simply put, gp120 draws the membrane bearing CD4 near the gp41 protein, and it is the gp41 protein that initiates the fusion reaction.

E. Association of gp120–gp41 with the Membrane

The gp41 protein is anchored to the membrane (Fig. 11). Mutations that remove or destroy the hydrophobic character of the second hydrophobic sequence of the gp41 protein result in release

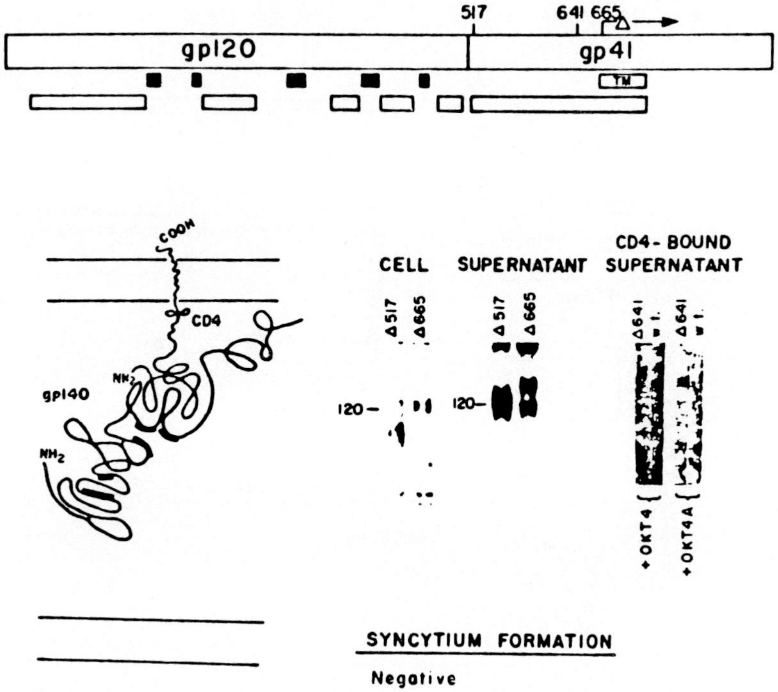

Fig. 11. Mutations within crucial hydrophobic domains of the HIV envelope gp41 result in loss of the ability of the transmembrane protein to anchor within the lipid bilayer. For abbreviations, see Fig. 7.

of the gp120–gp41 complex into the medium. No fusion reaction occurs. We propose that the second hydrophobic region of gp41 is the membrane-spanning region that serves as an anchor for the *env* protein to the membrane of infected cells and virions.

F. The Tail of gp41

One of the most unusual features of HIV is the long 150-amino-acid carboxy terminus of gp41 (Fig. 12). The corresponding region of many other retroviruses is only 12–15 amino acids in length. Deletion of much of this region does not destroy the ability of the

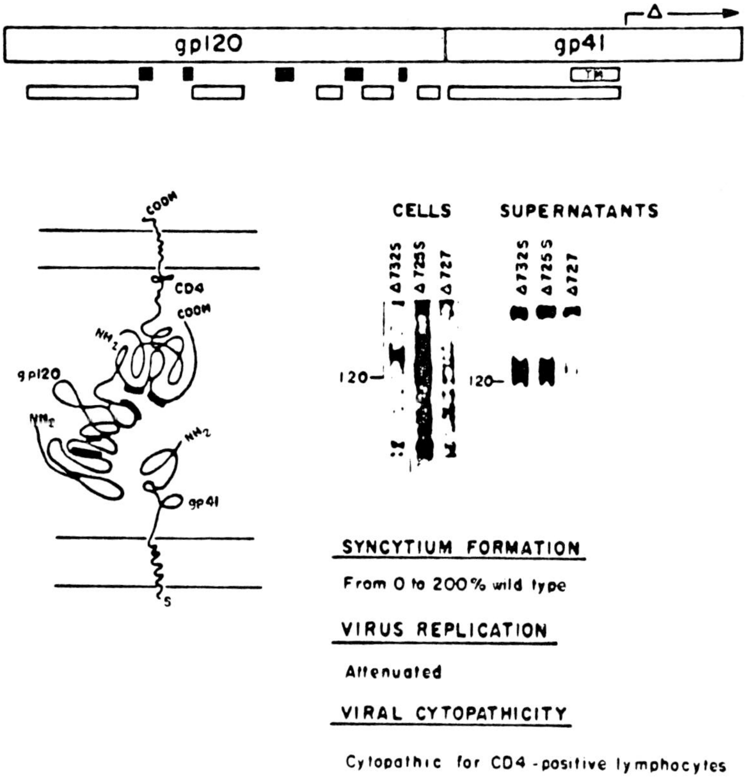

Fig. 12. Mutations within the carboxy terminus of the HIV envelope gp41 produce *env* products with normal processing, CD4 binding, and fusion activities. Viruses carrying such mutations are nevertheless severely attenuated in their replication. For abbreviations, see Fig. 7.

env protein to bind CD4 and to initiate fusion reactions as measured by cell–cell fusion, a reaction in which the *env* protein is located on the surface of an infected cell. Yet our previous studies demonstrated that deletion of 43 amino acids from the end of this region produced a virus that replicated much more slowly than did wild-type virus (Terwilliger *et al.*, 1986). We propose that the carboxy terminus of the gp41 is critical to the formation of infectious virus.

The model for envelope glycoprotein function provides insight into the details of a critical working component of HIV, one that both determines the type of cell infected and mediates CD4 cell killing. Such knowledge can be used to advantage for the creation of drugs and immunological reagents that inhibit cell functions such as CD4 binding and fusion reactions, that disrupt gp120–gp41 association, and/or that inhibit the cleavage of gp160.

G. SIV and HIV-2 (HTLV-IV) *env* Genes

In this regard we note that one of the key differences between HIV and STLV-III and some strains of HIV-2 (including HTLV-IV) is the presence of a stop codon located so as to produce a transmembrane protein that lacks the carboxy-terminal region of gp41 (Fig. 13) (Franchini *et al.*, 1987; Chakrabarti *et al.*, 1987; Guyader *et al.*, 1987). The open reading frame of gp41 encoding the carboxy terminus is present in these isolates save for the stop codon. Isolates of HIV-2 that produce both a long and short version of gp41 are also reported to exist. Some HIV-2 isolates are reported to replicate well in CD4 cells and to demonstrate a cytopathic effect, whereas others are reported to replicate poorly and are less cytopathic.

We propose that the structure of the carboxy-terminal region of the envelope glycoprotein of the simian and HIV-2 viruses is critical to the cytopathic effect of the virus both in cell culture and in animals. This region is proposed to affect infectivity of the cell-free virus.

We note that stop codon suppression is a common mechanism whereby retroviruses produce the *pol* genes. Such stop codon suppression may account for the presence of short and long versions of the transmembrane proteins produced by the same virus isolates. One possible explanation for the absence of disease induced by STLV-III in African Green monkeys and induction of disease in another monkey species, the macaque, may be related to

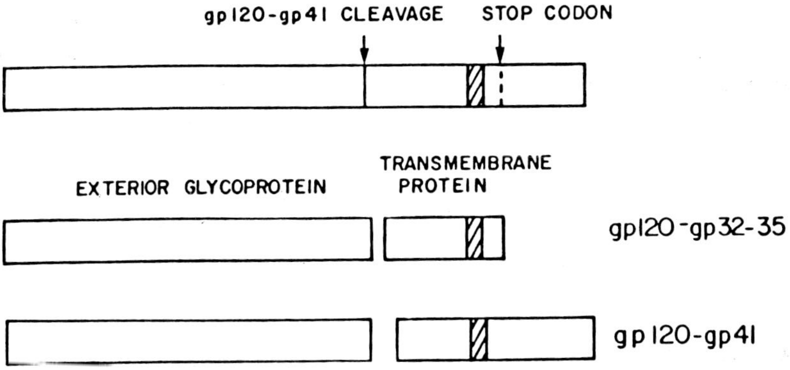

Fig. 13. Illustration of the stop codon present within the transmembrane protein coding sequences of SIV. Partial suppression of this codon may result in a mixture of two different types of transmembrane protein in infected cells.

relatively small differences in the efficiency of such stop codon suppression by the cells of these different species. If the suppression is more efficient in macaque cells than it is in African Green monkey cells, then more gp41 should be produced. The difference in gp41 levels should result in greater infectivity and cytopathic activity of the virus in macaques.

IV. Evasion of the Immune Response

Most people infected with HIV eventually mount a vigorous immune response, both humoral and cell-mediated, to HIV. Nonetheless, progressive disease usually ensues and there is some evidence that such immunity is not protective against new infections. Much of the antibody response to HIV is directed to the surface envelope glycoprotein and very high titers of anti-*env* antisera can be detected in the sera of many infected people (Barin *et al.*, 1985; Sarngadharan *et al.*, 1984). Moreover, the reactivities of anti-*env* antibodies are directed towards conserved regions of the protein, as most anti-*env* patient antibodies recognize all HIV strains (Matthews *et al.*, 1986). We have found that despite high antibody activity in most patient antisera, inhibition of the key reactions of CD4 binding and fusion as measured by syncytium formation is rarely observed. These two reactions are key to virus infectivity, cell–cell transmission, and CD4+ cell killing. Failure to significantly inhibit these reactions is likely to account for the lack of immune protection to HIV infection. Evidently, HIV is one of those microorganisms designed to evade the immune system. Evasion strategies may include the following possibilities.

A. The Hidden Binding Site

The critical molecular contacts required for CD4–gp120 binding may be recessed within the folds of the protein. Studies with other viruses show that it is rare for antibodies to recognize such recessed regions of protein. The CD4 binding regions are interspersed with hypervariable regions that probably occupy the surface of this region on the *env* protein.

B. The Sugar Coat

Much of the gp120 is heavily glycosylated (Robey *et al.*, 1985; Allan *et al.*, 1985a). Over half the molecular weight of this protein is accounted for by sugar residues. Such sugar residues are not recognized by antibodies, as the sugar residues on the HIV *env* protein are the same as those of normal surface glycoproteins. Such heavy glycosylation serves to mask most of the potentially immunogenic sites of the envelope glycoprotein.

C. Decoy

Two small conserved amino acid sequences comprise the immunodominant epitopes for all HIV envelope proteins (Palker *et al.*, 1987). Antibodies to these epitopes are only weakly neutralizing.

Strong immunological reactions to these conserved regions may mask immunological reactions to weaker but more important functional regions of the protein. Moreover, antibody binding to these regions may block such antibodies as may bind to more critical regions of the protein.

D. High-Affinity Binding

The affinity of the gp120 binding to CD4 is reported to be very high, about 4×10^{-9} (Laskey *et al.*, 1987). The affinity of most antibodies is much weaker than this. Hence, antibodies may only serve to delay but not prevent infection.

E. Cooperation

It is likely that the gp120 binding to CD4 is a cooperative reaction. The virus spike contains either two or three gp120–gp41 molecules. Binding of one of these to CD4 is likely to facilitate binding of others. Moreover, once a virion or infected cell has been placed in close juxtaposition with a membrane bearing CD4 so that binding to the envelope glycoprotein can occur, a cooperative reaction whereby other envelope proteins react with adjacent CD4 molecules is likely to occur. Antibodies may be sterically excluded from such structures as well.

F. Variation

The HIV genome is variable. It is possible that mutants arise within an infected individual at a frequency that is high enough that neutralizing antibody responses are circumvented by alteration of critical epitopes. However, this possibility remains to be proved despite demonstrated variability in the primary amino acid sequences of HIV *env* proteins.

V. Summary

Progress in molecular biology of HIV permits the beginning of an understanding of the pathogenic effects of infection. Controlled replication observed in infected people has its basis in an elaborate set of novel control genes. The ability of the virus to produce large amounts of infectious virions, in some instances, helps to explain the transmission of the virus from person to person.

The ability of the virus to kill CD4+ T helper cells can be explained by known functions of the envelope glycoprotein, namely, CD4 binding and fusion reactions. The ability to infect but not to kill other types of cells can also help us to understand the persistent virus production in infected people.

A detailed picture of the envelope protein function has emerged. This picture helps us to explain how the virus may evade the immune system. Systematic exploitation of the knowledge gained of the molecular biology of the virus and the functional regions of the envelope glycoprotein will speed the design of novel antiviral drugs and vaccines.

Acknowledgments[1]

Much of the work described herein was done at Dana-Farber Cancer Institute in the Laboratory of Biochemical Pharmacology by Joseph Sodroski and Craig Rosen. Ernest Terwilliger, Andrew Dayton, Wei Chun Goh, and Roberto Patarca also participated in the work. The work was supported by a contract from the State of Massachusetts, grants to J. S. (87) and C. R. (107) from the American Foundation for AIDS Research, and grants from the National Institutes of Health (AI24845) (CA42098). The work was made possible by gifts of HIV-infected cell lines and provirus DNA from the Laboratory of Tumor Cell Biology of the National Cancer Institute.

References

Allan, J. S., Coligan, J. E., Barin, F., McLane, M., Sodroski, J., Rosen, C., Haseltine, W., Lee, T., and Essex, M. (1985a). Major glycoprotein antigens

[1] This article also appears in Hanafusa, Pinter, and Pullman, "Retroviruses and Disease", Academic Press, San Diego, California, 1989.

that induce antibodies in AIDS patients are encoded by HTLV-III. *Science* **228,** 1091–1094.

Allan, J. S., Coligan, J. E., Lee, T. H., McLane, M. F., Kanki, P. S., Groopman, J. E., and Essex, M. (1985b). A new HTLV-III/LAV encoded antigen detected by antibodies from AIDS patients. *Science* **230,** 810–815.

Anya, S. K., Guo, C., Josephs, S. F., and Wong-Staal, F. (1985). *Trans*-activator gene of HTLV-III. *Science* **229,** 69–73.

Barin, F., McLane, M., Allan, J., and Lee, T. (1985). Virus envelope protein of HTLV-III represents major target antigen for antibodies in AIDS patients. *Science* **228,** 1094–1985.

Barre-Sinoussi, F., Chermann, J. C., Rey, F., Nugeyre, M. T., Chamaret, S., Gruest, J., Dauguet, C., Axler-Blin, C., Brun-Vezinet, F., Rouzious, C., Rozenbaum, W., and Montagnier, L. (1983). Isolation of a T-lymphotropic retrovirus from a patient at risk for acquired immune deficiency syndrome (AIDS). *Science* **220,** 868–871.

Brand, A. H., Breeden, L., Abraham, J., Sternglanz, R., and Nasmyth, K. (1985). Characterization of a "silencer" in yeast: A DNA sequence with properties opposite to those of a transcriptional enhancer. *Cell (Cambridge, Mass.)* **41,** 41–48.

Chakrabarti, L., Guyader, M., Alizon, M., Dantel, M., Desrosiers, R., Tiollais, P., and Sonigo, P. (1987). Sequence of simian imunodeficiency virus from macaque and its relationship to other human and simian retroviruses. *Nature (London)* **328,** 543–547.

Cheng-Meyer, C., Rutka, J. T., Rosenblum, M. L., Mchugh, T., Stittes, D. P., and Levy, J. A. (1987). Human immunodeficiency virus can productively infect cultured human glial cells. *Proc. Natl. Acad. Sci. U.S.A.* **84,** 3526–3530.

Chiodo, F., Fuerstenberg, S., Gidlund, M., Asjo, B., and Fenyo, E. M. (1987). Infection of brain-derived cells with the human immunodeficiency virus. *J. Virol.* **61,** 1244–1247.

Cullen, B. R. (1986). *Trans*-activation of human immunodeficiency virus occurs via a biomedial mechanism. *Cell (Cambridge, Mass.)* **46,** 973–982.

Dayton, A. I., Sodroski, J. G., Rosen, C. A., Goh, W. C., and Haseltine, W. A. (1986). The *trans*-activator gene of HTLV-III is required for replication. *Cell (Cambridge, Mass.)* **45,** 941–947.

Eyster, M. E., Goedert, J. J., Sarngadharan, M. G., Weiss, S. H., Gallo, R. C., and Blattner, W. A. (1985). Development and early natural history of HTLV-III antibodies in persons with hemophilia. *JAMA, J. Am. Med. Assoc.* **253,** 2219–2223.

Feinberg, M. B., Jarrett, R. F., Aldovini, A., Gallo, R. C., and Wong-Staal, F. (1986). HTLV-III expression and production involve complex regulation at the levels of splicing and translation of viral RNA. *Cell (Cambridge, Mass.)* **46,** 807–817.

Fisher, A. G., Ratner, L., Mitsuya, H., Marselle, L. M., Harper, M. E., Broder, S., Gallo, R. C., and Wong-Staal, F. (1986a). Infectious mutants of HTLV-III with changes in the 3′ region and markedly reduced cytopathic effects. *Science* **233**, 655–658.

Fisher, A. G., Feinberg, M. B., Josephs, S. F., Harper, M. E., Marselle, L. M., Reyes, G., Gonda, M. A., Aldovini, A., Debouk, C., Gallo, R. C., and Wong-Staal, F. (1986b). The trans-activator gene of HTLV-III is essential for virus replication. *Nature* (*London*) **320**, 367–371.

Fisher, A. G., Ensoli, B., Ivanoff, L., Chamberlain, M., Petteway, S., Ratner, L., Gallo, R. C., and Wong-Staal, F. (1987). The *sor* gene of HIV-1 is required for efficient virus transmission in vitro. *Science* **237**, 888–893.

Franchini, G., Gurgo, C., Guo, H. G., Gallo, R. C., Collati, E., Fergnoli, K. A., Hall, L. F., Wong-Staal, F., and Reitz, M. S. (1987). Sequence of simian immunodeficiency virus and its relationship to the human immunodeficiency virus. *Nature* (*London*) **328**, 539–543.

Franza, B. R., Josephs, S. F., Gilman, M. Z., Ryan, W., and Clarkson, B. (1987). Characterization of cellular proteins recognizing the HIV enhancer using a microscale DNA-affinity precipitation technique. *Nature* (*London*) **330**, 391–395.

Gartner, S., Markovits, P., Markovits, D., Kaplan, M., Gallo, R. C., and Popovic, M. (1986). The role of mononuclear phagocytes in HTLV-III/LAV infection. *Science* **233**, 215–219.

Goedert, J. J., Biggar, R. J., Winn, D. M., Greene, M. H., Mann, D. L., Gallo, R. C., Sarngadharan, M. G., Weiss, S. H., Grossman, R. J., Bodner, A. J., Strong, D. M., and Blattner, W. A. (1984). Determinants of retrovirus (HTLV-III) antibody and immunodeficiency conditions in homosexual men. *Lancet* **2**, 711–715.

Goedert, J. J., Biggar, R. J., Weiss, S. H., Eyster, M. E., Melbye, M., Wilson, S., Ginzburg, H. M., Grosman, R. J., Digioia, R. A., Sanchez, W. C., Giron, J. A., Ebbesen, P., Gallo, R. C., and Blattner, W. A. (1986). Three-year incidence of AIDS in five cohorts of HTLV-III-infected risk group members. *Science* **231**, 992–995.

Goh, W. C., Rosen, C., Sodroski, J. G., Ho, D. D., and Haseltine, W. A. (1986). Identification of a protein encoded by the *trans*-activator gene *tat* III of HTLV-III. *J. Virol.* **59**, 181–184.

Guy, B., Kieny, M. P., Riviere, Y., LePerch, C., Dott, K., Girard, M., Montagnier, L., and Lecocq, J. P. (1987). HIV F/3′*orf* encodes a phosphorylated GTP-binding protein resembling an oncogene product. *Nature* (*London*) **330**, 266–269.

Guyader, M., Emerman, M., Sonigo, P., Clavel, F., Montagnier, L., and Alizon, M. (1987). Genome organization and transactivation of the human immunodeficiency virus type 2. *Nature* (*London*) **326**, 662–669.

Hauber, J., Perkins, A., Heimer, E. P., and Cullen, B. R. (1987). Trans-activation of human immunodeficiency virus gene expression is mediated by nuclear events. *Proc. Natl. Acad. Sci. U.S.A.* **84**, 6364–6368.

Kao, S. Y., Galman, A. F., Luciw, P. A., and Peterlin, B. M. (1987). Antitermination of transcription within the long terminal repeat of HIV-1 by *tat* gene product. *Nature (London)* **330**, 489–493.

Klatzmann, D., Barre-Sinoussi, F., Nugeyre, M. T., Dauguet, C., Vilmer, E., Griscell, C., Brun-Vezinet, F., Rouzioux, C., Gluckman, J. C., Chermann, J. C., and Montagnier, L. (1984). Selective tropism of lymphadenopathy associated virus (LAV) for helper-inducer T lymphocytes. *Science* **225**, 59–63.

Knight, D. M., Flomerfelt, F. A., and Ghrayeb, J. (1987). Expression of the art/trs protein of HIV and study of its role in viral envelope synthesis. *Science* **236**, 837–840.

Kowalski, M., Potz, J., Basiripour, L., Dorfman, T., Goh, W. C., Terwilliger, E. F., Dayton, A., Rosen, C., Haseltine, W., and Sodroski, J. (1987). Functional regions of the envelope glycoprotein of human immunodeficiency virus type I. *Science* **237**, 1351–1355.

Laskey, L. A., Nakamura, G., Smith, D. H., Fennie, C., Shimasaki, C., Patzer, E., Berman, P., Gregory, T., and Capon, D. J. (1987). Delineation of a region of the human immunodeficiency virus type 1 gp120 glycoprotein critical for interaction with the CD4 receptor. *Cell (Cambridge, Mass.)* **50**, 975–985.

Luciw, P. A., Cheng-Mayer, C., and Levy, J. A. (1987). Mutational analysis of the human immunodeficiency virus: The orf-B region down regulates viral replication. *Proc. Natl. Acad. Sci. U.S.A.* **84**, 1434–1438.

McDougal, J. S., Kennedy, M. S., Sligh, J. M., Cart, S. P., Mawle, A., and Nicholson, J. K. A. (1986). Binding of HTLV-III/LAV to T4 + T cells by complex of the 110k viral protein and the T4 molecule. *Science* **231**, 382–385.

Matthews, T. J., Langlois, A. J., Robey, W. G., Chang, N. T., Gallo, R. C., Fischinger, P. J., and Bolognesi, D. P. (1986). Restricted neutralization of divergent human T-lymphotropic virus type III isolates by antibodies to the major envelope glycoprotein. *Proc. Natl. Acad. Sci. U.S.A.* **83**, 9709–9713.

Muesing, M. A., Smith, D. H., and Capon, D. J. (1987). Regulation of mRNA accumulation by a human immunodeficiency virus *trans*-activator protein. *Cell (Cambridge, Mass.)* **48**, 691–701.

Palker, T. J., Mathews, T. J., Clark, M. E., Ciancolo, G. J., Randall, R. R., Langlois, A. J., White, G. C., Sefei, B., Snyderman, R., Bolognesi, D. P., and Haynes, B. F. (1987). A conserved region at the COOH terminus of HIV gp120 envelope protein contains an immunodominant epitope. *Proc. Natl. Acad. Sci. U.S.A.* **84**, 2479.

Peterlin, B. M., Luciw, P. A., Barr, P. J., and Walker, M. D. (1986). Elevated levels of mRNA can account for the trans-activation of human immunodeficiency virus. *Proc. Natl. Acad. Sci. U.S.A.* **83**, 9734–9738.

Popovic, M., Saragadharan, M. G., Read, E., and Gallo, R. C. (1984). Detection, isolation and continuous production of cytopathic retrovirus (HTLV-III) from patients with AIDS and pre-AIDS. *Science* **244**, 497–500.

Robey, W. G., Safai, B., Oroszlan, S., Arthur, L. O., Gonda, M. A., Gallo, R. C., and Fischinger, P. J. (1985). Characterization of envelope and core structural gene products of HTLV-III with sera from AIDS patients. *Science* **228**, 593.

Rosen, C. A., Sodroski, J. G., and Haseltine, W. A. (1985). The location of cis-acting regulatory sequences in the human T cell lymphotropic virus type III (HTLV-III/LAV) long terminal repeat. *Cell (Cambridge, Mass.)* **41**, 813–823.

Rosen, C. A., Sodroski, J. G., Goh, W. C., Dayton, A. I., Lippke, J., and Haseltine, W. A. (1986a). Post-transcriptional regulation accounts for the *trans*-activation of HTLV-III. *Nature (London)* **319**, 555–559.

Rosen, C. A., Sodroski, J. G., Willems, L., Kettmann, R., Campbell, K., Zaya, R., Burny, A., and Haseltine, W. A. (1986b). The 3′ region of bovine leukemia virus genome encodes a *trans*-activator protein. *EMBO J.* **5** (10), 2585–2589.

Rosen, C. A., Terwilliger, E. F., Dayton, A. I., Sodroski, J. G., and Haseltine, W. A. (1988). Intragenic cis-acting *art* gene responsive sequences of the human immunodeficiency virus. *Proc. Natl. Acad. Sci. U.S.A.* **85**, 2071–2075.

Salahuddin, S. Z., Rose, R. M., Groopman, J. E., Markham, P. D., and Gallo, R. C. (1986). Human T lymphotropic virus type III infection of human alveolar macrophages. *Blood* **68**, 281–284.

Sarngadharan, M. G., Popovic, M., Bruch, L., Schupbach, J., and Gallo, R. C. (1984). Antibodies reactive with human T-lymphotropic retroviruses (HTLV-III) in the serum of patients with AIDS. *Science* **224**, 506–508.

Sodroski, J., Rosen, C., Wong-Staal, F., Salahudding, S. Z., Popovic, M., Anya, S., Gallo, R. C., and Haseltine, W. A. (1985a). *Trans*-acting transcriptional regulation of HTLV-III long terminal repeat. *Science* **227**, 171–173.

Sodroski, J., Patarca, R., Rosen, C., and Haseltine, W. A. (1985b). Location of the *trans*-activating region on the genome of HTLV-III. *Science* **229**, 74–77.

Sodroski, J. G., Goh, W. C., Rosen, C., Dayton, A., Terwilliger, E., and Haseltine, W. (1986a). A second post-transcriptional *trans*-activator gene required for HTLV-III replication. *Nature (London)* **321**, 412–417.

Sodroski, J., Goh, W. C., Rosen, C., Tartar, A., Portelle, D., Burny, A., and Haseltine, W. (1986b). Replicative and cytopathic potential of HTLV-III/LAV with *sor* gene deletions. *Science* **231**, 1549–1553.

Starcich, B., Ratner, L., Josephs, S. F., Okamoto, T., Gallo, R. C., and Wong-Staal, F. (1985). Characterization of long terminal repeat sequences of HTLV-III. *Science* **227**, 538–540.

Strebel, K., Dougherty, D., Clouse, K., Cohen, D., Folks, T., and Martin, M. (1987). The HIV "A" (sor) gene product is essential for virus infectivity. *Nature* (*London*) **328**, 728–730.

Terwilliger, E. F., Sodroski, J. G., Rosen, C. A., and Haseltine, W. A. (1986). Effects of mutations within the 3′*orf* open reading frame region of HTLV-III/LAV on replication and cytopathogenicity. *J. Virol.* **60**, 754–760.

Terwilliger, E. F., Sodroski, J. G., Haseltine, W. A., and Rosen, C. R. (1988). The *art* protein of HIV is essential for virus replication. *J. Virol.* **62**, 655–658.

Wright, C. M., Felber, B. K., Paskolis, H., and Paulakis, G. N. (1986). Expression and characterization of the *trans*-activator of HTLV-III/LAV virus. *Science* **234**, 988–992.

12

Perturbation of DNA Synthesis and the Generation of Drug Resistance in Cultured Mammalian Cells

ROBERT T. SCHIMKE, JEFF L. ELLSWORTH, CYNTHIA HOY, R. IVAN SCHUMACHER, AND STEVEN W. SHERWOOD

Department of Biological Sciences
Stanford University
Stanford, California

I. Introduction

We have been studying the process whereby cultured animal cells become resistant to methotrexate (MTX) as a result of amplification of the dihydrofolate reductase (DHFR) gene (for reviews of gene amplification, see Schimke, 1984; Stark and Wahl, 1984; Hamlin *et al.*, 1984). We have found that a variety of treatments of cells can increase the frequency of MTX resistance and gene amplification, including drugs that inhibit DNA synthesis such as hydroxyurea (Brown *et al.*, 1983) and aphidicolin (Johnston *et al.*, 1986; Hoy *et al.*, 1987); agents that interact with DNA, including carcinogens and UV radiation (Tlsty *et al.*, 1984; Hoy *et al.*, 1987); and hypoxia (Rice *et al.*, 1986). All such treatments have in common the inhibition of DNA synthesis, and resumption of DNA synthesis is necessary for the generation of the subset of cells in which the frequency of MTX resistance is increased during subsequent drug selection.

During the past 2 years we have made extensive use of a flow cytometric technique that permits determining the fate of cells based on their position in the cell cycle prior to treatment (inhibition of DNA synthesis). The method employs the use of a fluoresceinated antibody directed against BrdU-substituted DNA to identify S-phase cells, plus propidium iodide to measure total cell DNA content. Figure 1 shows a representative experiment, in which an asynchronous population of Chinese hamster ovary (CHO) cells was treated with aphidicolin (Hoy *et al.*, 1987). Figure 1 (0 hr) shows cells labeled with BrdU for 20 min and then subjected to 18 hr of treatment with aphidicolin. Those cells which were in G_2–M at the time of BrdU labeling are now present as G_1 cells (by DNA content measurement and lack of anti-BrdU fluorescence). The remaining panels of Fig. 1 indicate the progression of cells at the indicated hours following restoration of DNA synthesis. Those cells which were in either G_1 or G_2 and M during the 20-min labeling period (showing no incorporation of BrdU) progressed through the cell cycle, as shown by variation in the number of cells in different cell cycle compartments on the horizontal axis (DNA). Compared to normal cells, these G_1-arrested cells behaved differently from cells progressing through a normal S phase, in that the time required to progress from G_1 back into G_1 was shortened from the normally occurring 15 hr to as short a time as 6–8 hr (see Hoy *et al.*, 1987). The cells that were in S phase at the time of inhibition of DNA synthesis [those cells with fluorescence above background (vertical axis of Fig. 1)] behaved in

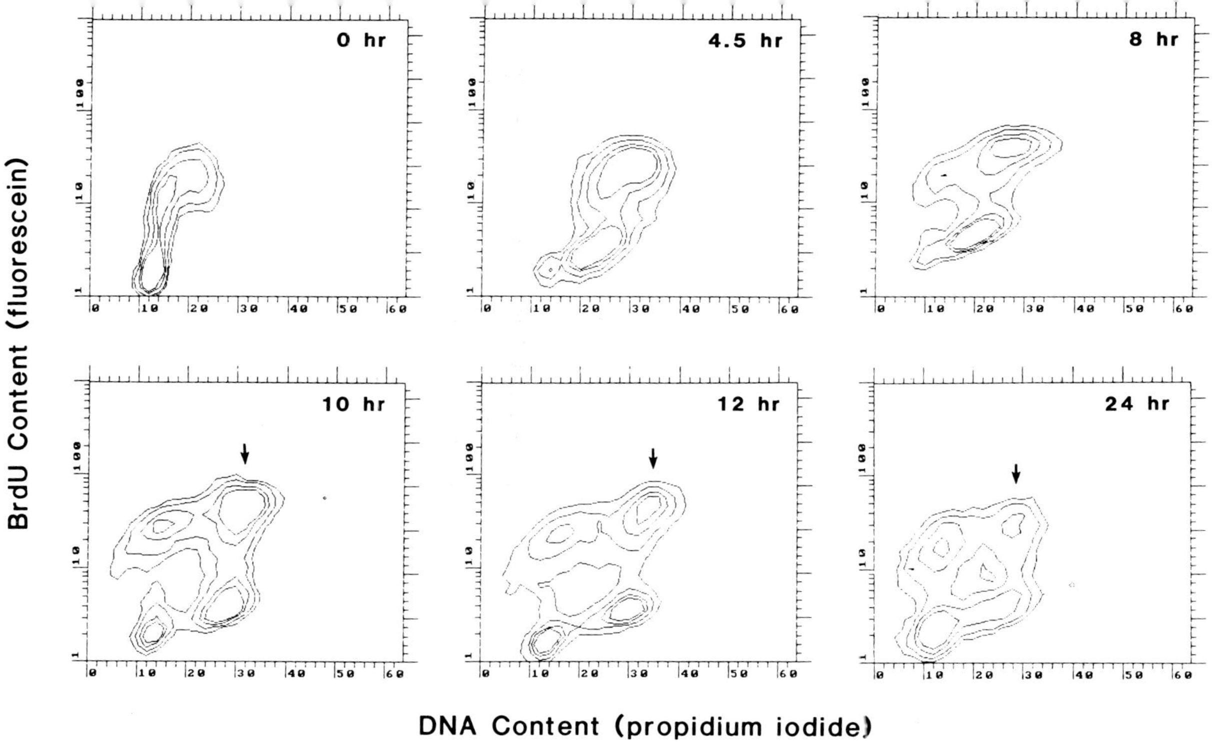

Fig. 1. Contour plots of bivariate DNA/BrdU distributions for Chinese hamster ovary cells blocked with 6 μM aphidicolin after 20 min labeling with BrdU. The six panels show plots of cells chased in BrdU-free medium after removal of aphidicolin from the media for the times indicated. Arrows indicate the border for normal G_2–M (4C) DNA content per cell, and cells with more than this DNA quantity have "overreplicated" DNA prior to mitosis (see Hoy *et al.*, 1987, for details).

a dramatically different fashion. As these cells progressed through the cell cycle, a population of cells with more than 4C DNA content (arrow indicates boundary of normal 4C DNA content) was generated. This is clearly different from those cells arrested in G_1, which do not progress beyond the limit of a 4C DNA content per cell. Additionally, a large proportion of such cells had a delay in the time they go through mitosis (see Hoy *et al.,* 1987), i.e., mitosis occurred 18–36 hr after restoration of DNA synthesis. We have found this same pattern with cells treated with hydroxyurea or UV radiation (Hoy *et al.,* 1987) or hypoxia (Rice *et al.,* 1986) as well.

Figure 2 shows another general finding of our results, i.e., that the increased frequency of MTX resistance as generated by such treatments comes from that subset of a cell population that has >4C DNA content (Hill and Schimke, 1985; Rice *et al.,* 1986; Johnston *et al.,* 1986). In Fig. 2 cells were stained with Hoechst 33342 and sorted as indicated by the brackets (inset). Plating into MTX shows that it was the cells with >4C DNA content that had the higher frequency of resistance (see Hoy *et al.,* 1987, for details).

II. The Mechanism of Gene Amplification

We have favored the concept that the initial step in the generation of additional copies of a gene (gene amplification) results from aberrant DNA synthesis, i.e., overreplication (see Schimke *et al.,* 1986), and the flow cytometric data (see above) is consistent with such a general concept. One mechanism for the generation of additional DNA per cell is as a result of additional initiations of DNA replication in a single, normally progressing cell cycle (see Fig. 3), and we have provided data consistent with such a mechanism (Mariani and Schimke, 1984). Our current results (see above) showing that the M phase of the cell cycle is delayed suggests another mechanism, one in which the normally occurring mitosis (M) is delayed relative to S phase, and occurs variably

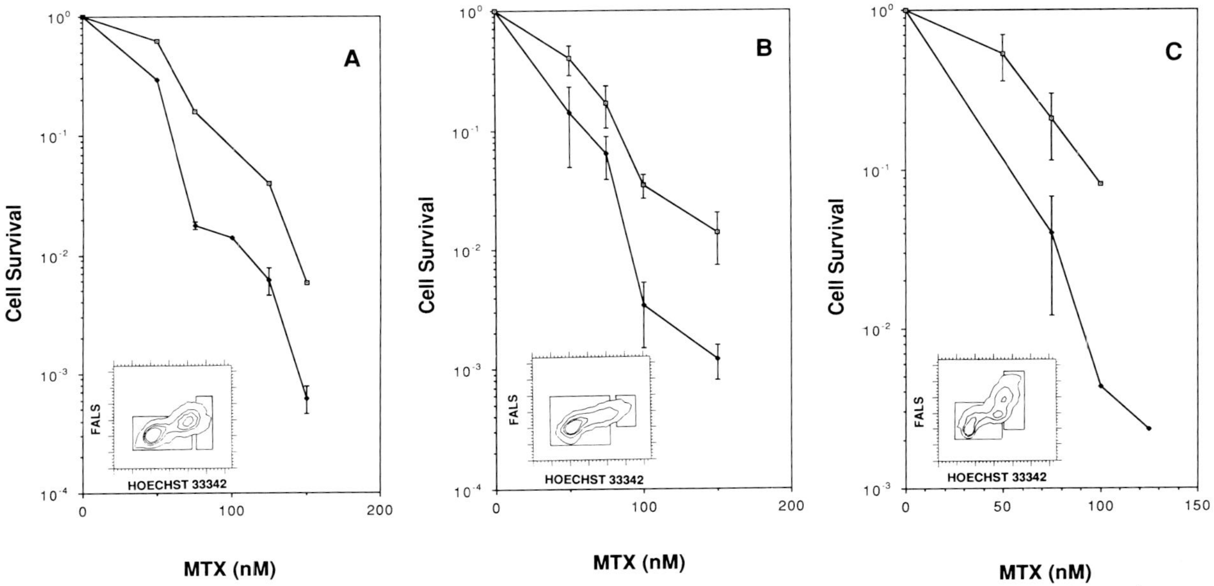

Fig. 2. Survival of cells sorted containing high or low DNA content in methotrexate (0–150 n*M*). Chinese hamster ovary cells were stained with Hoechst 33342 after treatment with 12 J/m^2 UV (A), 1 m*M* hydroxyurea (B), or 1 μg/ml of aphidicolin (C). Contour plots of Hoechst-stained cells and windows for high and low DNA content from which cells were sorted are shown in the insets of each of the panels. Forward angle light-scatter (FALS) is plotted against Hoechst fluorescence (see Hoy *et al.*, 1987, for details).

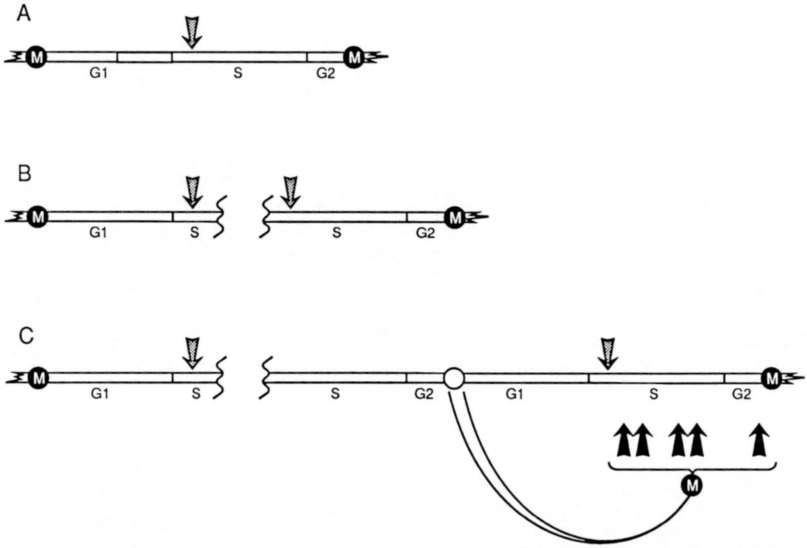

Fig. 3. Theoretical representations of two mechanisms for the generation of cells with greater than normal DNA content/cell prior to mitosis. (A) represents a normal cell cycle. The arrow indicates the time of replication of a specific gene. (B) represents the situation in which there are two replications of a specific gene and the normal relationship between S and M is retained, i.e., M occurs following S. The broken line indicates inhibition of DNA synthesis. In this mechanism there is an additional initiation of replication of DNA following restoration of DNA synthesis but prior to a mitosis whose timing relative to S phase is as in unperturbed cells. (C) depicts the situation in which inhibition of DNA synthesis does not alter the timing of replication of the specific gene in an S phase. Rather, inhibition of DNA synthesis results in a delay in mitosis such that it can occur at varying times in a second cell cycle (as defined by S to S).

within a "second" S-phase (see Fig. 3). From the standpoint of alterations of normal regulatory processes, the regulation of initiation of replication would be altered in the first mechanism, whereas the "lesion" in the second mechanism would relate to the proper integration of cell cycle events. These two mechanisms are not mutually exclusive and may occur under different experimental conditions in different cell types. The concept that cell cycle progression and the dissociation of S phase and M phases of cell cycles can be perturbed has been suggested previously (Ehmann *et al.*, 1975; Merz and Schneider, 1983; Schlegel and Pardee, 1987).

III. Chromosomal Aberrations following Inhibition of DNA Synthesis

One of the striking effects of treatment with the above agents is the generation of cells with a variety of chromosomal aberrations. In an extensive study, Hill and Schimke (1985) treated mouse L5178Y cells with hydroxyurea, and studied cells as they underwent the first mitosis following inhibition of DNA synthesis. They observed that the first cells undergoing mitosis after hydroxyurea inhibition contained no chromosomal aberrations, whereas cells undergoing mitosis 8–18 hr after resumption of DNA synthesis contained a variety of defects, including normal chromosomes with extrachromosomal DNA, varying degress of partial ploidization (often with partially condensed chromosomes), and variously gapped, broken, or fragmented chromosomes, as well as occasional metaphases with complete endoreduplication of chromosomes. All of the cells with chromosomal aberrations were derived from that subset of cells with >4C DNA content. With our current data we can begin to understand why chromosomal defects were observed only in those cells whose mitosis occurred at a later time following recovery of DNA synthesis. As shown in Fig. 1, the first cells to undergo mitosis were actually those arrested in G_1, whereas it is the cells that are arrested in S phase that are delayed in M, and that have >4C DNA content prior to mitosis.

Figure 4 shows three representative types of chromosomal aberrations characteristic of cells undergoing delayed mitosis following inhibition of DNA synthesis. Panel A shows normally condensed chromosomes with extrachromosomal DNA. Panel B shows a metaphase spread with partial ploidization in which the individual chromosomes are in differing states of condensation. Panel C shows the most frequent type of chromosomal abnormal-

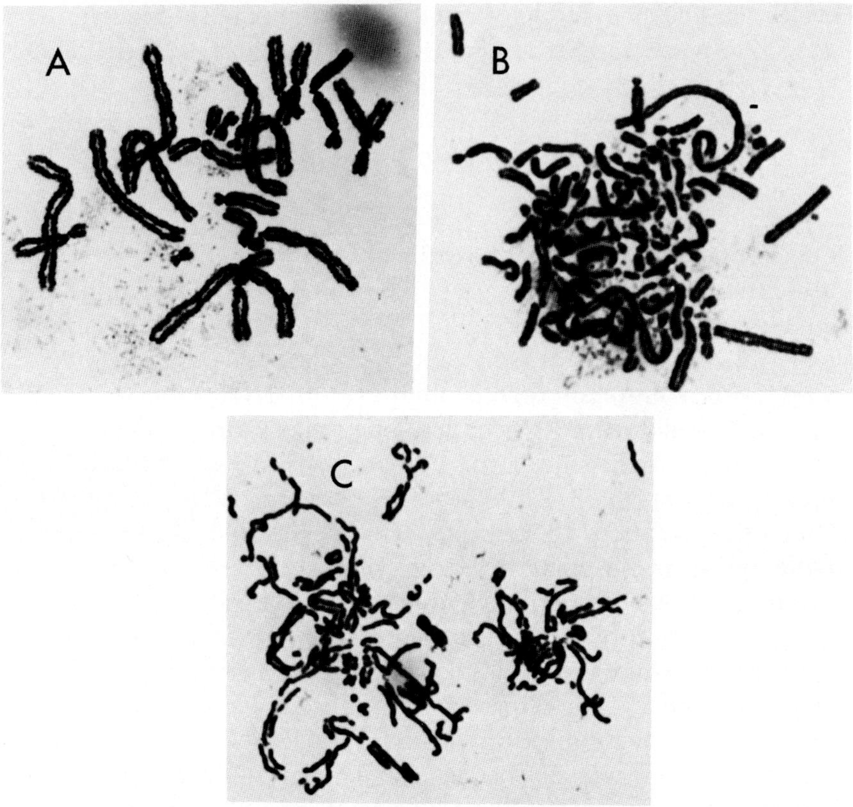

Fig. 4. Examples of metaphase chromosome spreads of Chinese hamster ovary cells observed in the first mitosis occurring after recovery of inhibition of DNA synthesis by aphidicolin. (A) A metaphase spread showing condensed chromosomes and extrachromosomal DNA; (B) a metaphase showing a polyploidy in which the chromosomes are variable in the degree of their condensation; (C) a metaphase spread showing extensive chromosomal breaks and gaps.

ity, a metaphase spread with extensively fragmented chromosomes. Clearly, cells with extensive chromosomal fragmentation are unlikely to survive, and we attribute the cell death following inhibition of DNA synthesis to the generation of cells with broken chromosomes. In fact, we observe that cells do not die, i.e., lose attachment to plastic and become "floaters," in the presence of DNA synthesis inhibitors for up to 36–48 hr of treatment. They "die" only after the drugs are removed, cell cycle progression resumes, and mitosis commences. Interestingly, Sherwood *et al.* (1988) found that inhibition of protein synthesis during the time of inhibition of DNA synthesis reduced cell killing and chromosomal abnormalities, and prevented the enhancement of MTX resistance. Thus the killing of cells and the enhanced resistance to drugs appear to be active responses of cells to the inhibition of DNA synthesis.

IV. The Frequency of Chromosomal Abnormalities in Normally Dividing Cells

We have proposed (Schimke *et al.*, 1986) that perturbation of DNA synthesis patterns in cultured cells can result in "overreplication" of DNA (see Fig. 3), among the consequences of which are the generation of broken or fragmented chromosomes (cell death), as well as the more rare recombinational events which can result in gene amplification and deletion events. If such events have selective advantages, such as growth in the presence of cancer chemotherapeutic agents (drug resistance) or the capability to overcome normal growth constraints *in vivo* (cancer), such rare events will overshadow what is obviously the common consequence, i.e., cell death.

An important question to be considered is whether the cell death–fragmented chromosome phenomenon studied in cultured cells has any counterpart *in vivo* in rapidly proliferating cell types. We are currently exploring this question using the small intestine of the intact rat. In this system the villus cells undergo synchronous mitosis, resulting in replacement of the entire villus epithelium every 48 hr (Leblond and Stevens, 1948). Table I provides the results of experiments in which rat jejunal crypt cell metaphases were examined 1.5 hr following administration of colchicine to intact rats between 8 and 9 a.m. The type of chromosomal

TABLE I

Frequency of Chromosome Aberrations in Jejunal Crypt Cells in Young and Aged Fischer 344 Rats Obtained from the National Institute of Aging Rat Colony[a]

Rat age (months)	Altered metaphases (No.) Breaks	Gaps	Fragments	Pulverized	Metaphases with more than one type of damage (%)
3–7	4	42	13	8	8
24–26	17	71	16	15	13

[a] For the 3–7-month- and 24–26-month-old groups, 164 ± 24 and 150 ± 25 (mean ± SD) metaphases per animal were analyzed, respectively. There were six animals in each group. A total of 982 and 904 metaphases were analyzed for the young and aged rats, respectively. Breaks are defined as a discontinuity of a single chromatid or symmetrically involving both sister chromatids in which there is misalignment of the proximal and distal segments and/or an achromatic region that is larger than the chromatid width. A gap is defined as unpaired and/or paired acentric fragments generally smaller than the chromatid width and chromatids lacking centromeric constrictions. A pulverized metaphase has virtually no chromosome structure and consists of finely scattered chromosome pieces of the same size as the chromatic granules of interphase nuclei.

abnormalities that were observed in young and old rats were essentially similar. Figure 5 shows two typical metaphase abnormalities. Figure 5A shows chromosomes that are extensively pulverized. Figure 5B shows a metaphase spread in which specific chromosome gaps are evident. In young rats the mean percentage of "damaged" metaphases was 6.2%, whereas in older rats the mean percentage was 11.6%. Table II shows the variability from animal to animal. In cases in which specific chromosomal breaks can be observed, we found that they occurred with a high frequency in specific chromosomes. We do not have any information on whether such breaks occur at the site(s) of specific genes. An additional parameter we are exploring is starvation. Witholding food from the rats for 2–3 days greatly reduces the cycling of

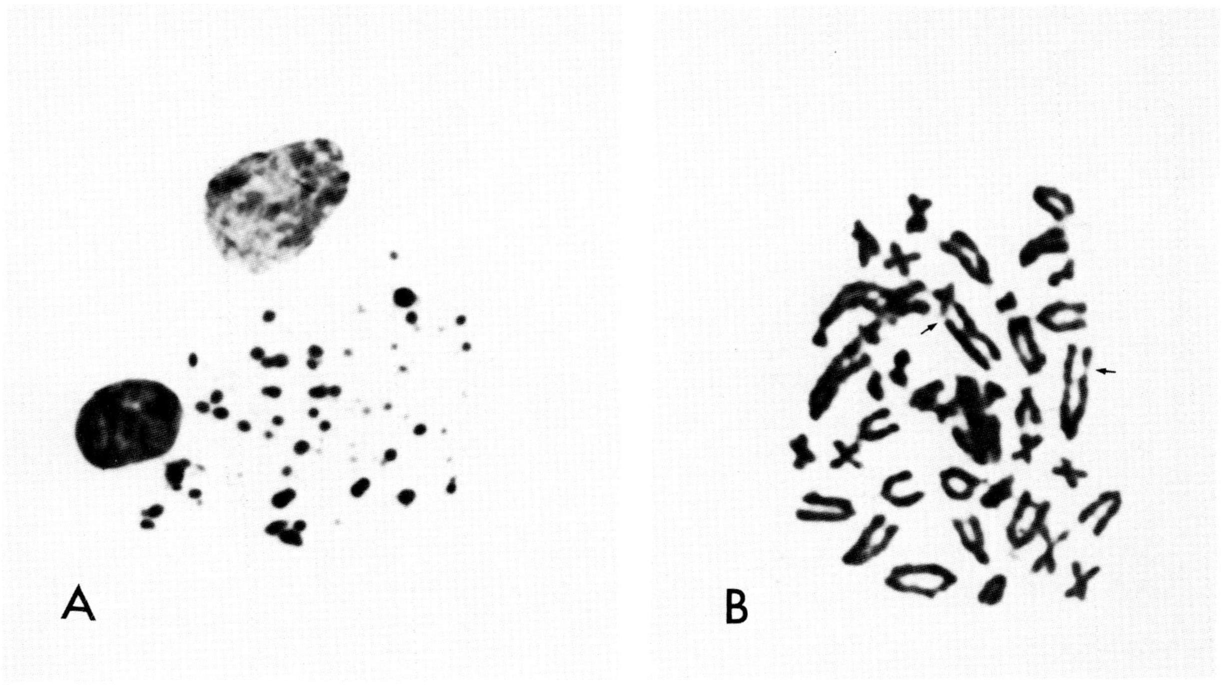

Fig. 5. Two representative abnormal metaphase spreads from rat jejunal crypt cells. (A) A pulverized metaphase; (B) a metaphase spread with chromosome breaks and gaps (arrows).

TABLE II

Summary of Metaphase Chromosome Damage in Young and Aged Rat Jejunal Crypt Cells[a]

Rat age (months)	Damaged metaphases[b] (% of total)	
3	5.5	
3	9.3	
4	7.0	
7	4.0	$X = 6.2 \pm 1.9\%$
7	4.7	
7	6.4	
24	17.6	
24	15.1	
25	5.6	$X = 11.6 \pm 4.4\%$
25	12.8	
26	8.4	
26	10.2	

[a] Data are listed by individual rat.
[b] Difference between groups significant ($p < 0.05$) by the Rank Sum Test. The values represent the mean ± SD.

crypt cells (Oliver-Brown *et al.*, 1963). When metaphase chromosomes were prepared from jejunal crypt cells from rats fasted for 3 days, we observed that the number of damaged metaphases increased approximately 2-fold in both young and old rats, to approximately 10 and 20% of metaphases, respectively (data not shown). These results suggest that chromosome "damage" (and attendant cell death) does occur in proliferating somatic cells and that it occurs variably as a function of age, as well as of a number of other physiological and individual parameters. It is surprising that it may involve as many as 20% of dividing cells. We are currently extending these types of study to other rapidly proliferating cell types in the intact animal.

V. Discussion

Our results indicate that a variety of agents, which have in common the inhibition of DNA synthesis, can enhance frequencies of MTX drug resistance and gene amplification. These same agents also kill cells. Our results are consistent with the proposal that the death of cells is at least in part a consequence of chromosomal breakage phenomena, and that this breakage is a result of cellular responses to inhibition and recovery of DNA synthesis. This process requires protein synthesis during the time of inhibition of DNA synthesis. Our analyses of progression of cells through the cell cycle following recovery of DNA synthesis indicates that the cells in which chromosomal breakage–damage phenomena occur are those cells whose DNA synthesis is inhibited when they are in S phase and that the mitosis that occurs following recovery is delayed variably during the time when a second S phase is occurring (i.e., an S phase that occurs following the completion of an S phase which was interrupted by inhibition of DNA synthesis; see Fig. 3). Such a dissociation of S and M phases of a cell cycle is associated with chromosomal breakage and cell death. However, this dissociation phenomenon is also associated with recombination–rearrangements resulting in increased frequencies of drug resistance. Thus, some of the commonly employed cancer chemotherapeutic agents whose use is based on their capacity to kill cells, may, as part of the killing process, facilitate the class of mutational events that results in the all-too-common phenomenon of drug resistance. The concept that events of the cell cycle can be dissociated is supported by the the cdc2 mutation in *Schizosaccharomyces pombe,* which can prevent cell cycle progression from G_1 into S or from G_2 to M. This mutant has a protein kinase C counterpart in the human genome (Lee and Nurse, 1987), perhaps suggesting that some aspect of protein phosphorylation as it affects either enzyme function or nuclear cytoarchitecture is important in such cellular decision making.

If the generation of cells with broken or fragmented chromosomes is an indication of the perturbation of the normal relationship between the S phase of a cell cycle and the subsequent mitosis, then perhaps our findings of a significant number of normally proliferating cells in the small intestine with damaged chromosomes is surprising. Inasmuch as such cells will not divide again, their presence will go undetected. If this phenomenon occurs with other rapidly proliferating cell types, in particular after mitogenic stimulus, then one might suggest that the substratum for rearrangement mutations (deletions and amplifications) that can result in loss of growth constraints, i.e., cancer, is being generated continually. Interestingly, in those intestinal cells in which limited chromatid breaks are detected, the breaks are apparently specific to certain chromosomes, i.e., they are not random (Ellsworth and Schimke, 1989). We are currently attempt ing to determine if these breaks are at sites of highly expressed genes. If so, the frequency and specificity for chromosome breakage may assist in an understanding of specific breakage–recombination events that underlie the translocation–activation phenomena of various types of leukemias.

VI. Summary

When DNA synthesis in an asynchronous population of cells is inhibited by various treatments, including hydroxyurea, aphidicolin, or UV irradiation, upon recovery, that subset of cells which were in S phase (S) at the time of inhibition have a delayed mitosis and contain greater than 4C DNA content per cell prior to mitosis (M). It is from this subset of the cell population that cells with increased frequencies of resistance to methotrexate are derived, as well as cells with major chromosomal aberrations, which are basically "dead" cells. We discuss the possible mechanisms involved both in the cell killing that results from perturbation of DNA synthesis and in increased frequencies of drug resistance in terms of perturbations of the normal progression from S phase into M phase of the cell cycle.

References

Brown, P. C., Tlsty, T. D., and Schimke, R. T. (1983). Enhancement of methotrexate resistance and dihydrofolate reductase gene amplification by treatment of mouse 3T6 cells with hydroxyurea. *Mol. Cell. Biol.* **3**, 1097–1107.

Ehmann, U. K., Williams, J. R., Nagel, W. A., Brown, J. A., Belli, J. A., and Lett, J. T. (1975). Perturbations in cell cycle progression from radioactive DNA precursors. *Nature (London)* **258**, 633–636.

Ellsworth, J., and Schimke, R. T. (1989). On the frequency of damaged metaphase chromosomes in rat intestinal crypt cells as a function of age. In preparation.

Hamlin, J. L., Milbrandt, J. D., Heintz, H. H., and Azizkhan, J. C. (1984). DNA sequence amplification in mammalian cells. *Int. Rev. Cytol.* **90**, 31–82.

Hill, A. B., and Schimke, R. T. (1985). Increased gene amplification in L 5178Y mouse lymphoma cells with hydroxyurea-induced chromosomal aberrations. *Cancer Res.* **45**, 5050–5057.

Hoy, C. A., Rice, G. C., Kovacs, M., and Schimke, R. T. (1987). Overreplication of DNA in S-phase CHO cells after DNA synthesis inhibition. *J. Biol. Chem.* **262**, 11927–11943.

Johnston, R. N., Feder, J., Hill, A. B., Sherwood, S. W., and Schimke, R. T. (1986). Transient inhibition of DNA synthesis results in increased dihydrofolate reductase synthesis and subsequent increased DNA content per cell. *Mol. Cell. Biol.* **6**, 3373–3381.

Leblond, C. P., and Stevens, C. E. (1948). Constant renewal of the intestinal epithelium in the albino rat. *Anat. Rec.* **100**, 357–371.

Lee, M. G., and Nurse, P. (1987). Complementation used to clone a human homologue of the fission yeast cell cycle control gene cdc28. *Nature (London)* **322**, 311–335.

Mariani, B. D., and Schimke, R. T. (1984). Gene amplification in a single cell cycle in Chinese hamster ovary cells. *J. Biol. Chem.* **259**, 1901–1910.

Merz, R., and Schneider, R. (1983). Growth characteristics of anaerobically treated early and late s-period of Ehrlich ascites tumor cells after reaeration. *Z. Naturforsch.* **38**, 313–318.

Oliver-Brown, H., Levine, M. L., and Lipkin, M. (1963). Inhibition of intestinal epithelial cell renewal and migration induced by starvation. *Am. J. Physiol.* **205**, 868–872.

Rice, G. C., Hoy, C., and Schimke, R. T. (1986). Transient hypoxia enhances the frequency of dihydrofolate reductase gene amplification in Chinese hamster ovary cells. *Proc. Natl. Acad. Sci. U.S.A.* **83**, 5978–5982.

Schimke, R. T. (1984). Gene amplification in cultured animal cells. *Cell (Cambridge, Mass.)* **37**, 705–713.

Schimke, R. T., Sherwood, S. W., Hill, A. B., and Johnston, R. N. (1986). Overreplication and recombination of DNA in higher eukaryotes: Potential consequences and biological implications. *Proc. Natl. Acad. Sci. U.S.A.* **83**, 2157–2161.

Schlegel, R., and Pardee, A. B. (1987). Periodic mitotic events induced in the absence of DNA replication. *Proc. Natl. Acad. Sci. U.S.A.* **84**, 9025–9029.

Sherwood, S. W., Schumacher, I., and Schimke, R. T. (1988). The effect of cycloheximide on the development of methotrexate resistance in Chinese hamster ovary cells treated with inhibitors of DNA synthesis. *Mol. Cell. Biol.* **8**, 2822–2827.

Stark, G. R., and Wahl, G. M. (1984). Gene amplification. *Annu. Rev. Biochem.* **53**, 447–491.

Tlsty, T. D., Brown, P. C., and Schimke, R. T. (1984). UV radiation facilitates methotrexate resistance and amplification of the dihydrofolate reductase gene. *Mol. Cell. Biol.* **4**, 1050–1056.

Index